MP³ MATHEMATICS
Post-Secondary Preparedness Package

Jack Weiner and Hoshang Pesotan
University of Guelph

MP³: <u>M</u>athematics <u>P</u>ost-Secondary <u>P</u>reparedness <u>P</u>ackage

by *Jack Weiner and Hosh Pesotan*
with *Sylvia Nguyen and George Hutchinson*

Interior design: Paul Futhey
Cover design: Zohra Karim

For more information contact:
Studentawards Inc.
603-110 Eglinton Ave East
Toronto, Ontario
M4P 2Y1
(416) 322-3210

www.studentawards.com

ISBN: 978-0-9812753-0-7

Get a head start on your math studies with Maple™!

Take advantage of Maplesoft™ math resources designed exclusively for students using MP³.
Go to: **www.maplesoft.com/mp3**

The Mathematics Post-Secondary Preparedness Package (MP³) is designed to help you prepare for courses that require an understanding of high school math concepts. Maple, math software created by Maplesoft, can help you get prepared as well!

Maple is a powerful general purpose math tool designed to provide an environment for students to explore and 'do' math. Used creatively, Maple can help students learn better and faster. It can illuminate theory, clarify the abstract and give form and substance to general principles.

Bring math concepts to life!

Math education has never been more important than it is today. Whether you are continuing your studies or going out into the world, understanding mathematics is an essential part of your success in this increasingly technological world. As a student, your challenge is to find ways to improve your understanding and increase your test scores.

Maple can help you meet these challenges!

Maple is used extensively in many mathematics courses at colleges and universities around the world. By gaining experience with Maple now you'll gain valuable experience for your post-secondary studies in the future. By using Maple in combination with MP³, in addition to understanding important high school mathematics concepts, you'll gain a solid understanding of the math tool used by academic institutions everywhere.

Discover for yourself why Maple is the math software tool of choice for students everywhere.

Go to: **www.maplesoft.com/mp3**

*In addition to Maple, Maplesoft also offers the PreCalculus Study Guide and the Calculus Study Guide as well as the Mathematics Survival Kit, an ebook written by Professor Jack Weiner, one of the authors of MP³. Learn more at **www.maplesoft.com/mp3***

Maplesoft
Mathematics • Modeling • Simulation

www.maplesoft.com | info@maplesoft.com

© Maplesoft, a division of Waterloo Maple Inc., 2009. Maplesoft, Maple, and MapleSim are trademarks of Waterloo Maple Inc. All other trademarks are the property of their respective owners.

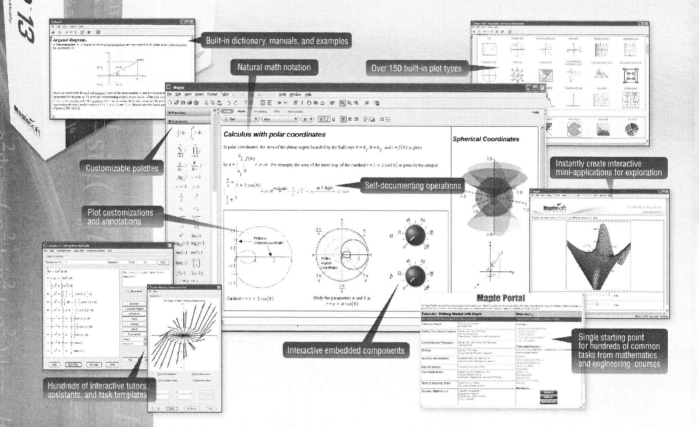

TABLE OF CONTENTS

Introduction

	Problems	Solutions

MP³

Welcome to students & Introduction to Problem Sets

Introducing the MP³: Mathematics
Post-Secondary Preparedness Package

Dear Jennifer, Ahmed, John, Peter, Emily, Suri, George, …, whoever you are!

Here is a sad truth from a veteran math professor. One of the biggest fears of many students entering university or college is that they will struggle because their math skills are so rusty. Math anxiety is rampant. Even many students whose high school math marks were ^{high} come to class worried.

Many should be worried. Some math profs, or in fact, profs teaching ANY course that uses or builds on math, will just assume you have your high school math down pat and at your fingertips.

You may not. You shut your books a while ago and haven't solved a quadratic equation or found the hypotenuse of a right triangle or $\sin(30°)$ since **way back when!** Your basic math skills and knowledge are surrounded by neural cobwebs!

Have we got a deal for you: the Math Post-Secondary Preparedness Package! **MP³!**

Here are 9 problems sets dealing with topics from "Basic Numeracy" to "Exponents and Logarithms".

> *We did not include calculus since so many of you will not have taken a high school calculus course. Also, virtually every university first year calculus course will cover all the calculus of a high school course, but usually in more depth. However, we did include references to some trig formulas, for example, that you may not see "officially" till you take a calculus course. We included logs and exponents, including base "e", the "natural log base". Again, you may not see this formally till your calculus course.*

This is a FINITE package.

This is a DO-ABLE package.

We are not sending you a huge book or asking you to take an entire course. We want a reasonable amount of your time and concentrated effort.

As you read this, you likely have time before you begin your post-secondary adventure. You have time on your hands. Give us, give yourself, **one to two hours** a day for the next nine days. Clear the math mental cobwebs by working through these problem sets.

> *Bonus: this will prove a tremendously useful review package to keep handy as you progress through your program.*

Then, when that first crucial week of classes arrives all too soon, you will NOT get stuck in class on new math or physics or chemistry or… because the prof assumed and went too fast through some forgotten high school math.

9 problems sets: two hours max per day for the next week or two. Fair trade to ensure success!

How will you know your answers are right? After the problem sets, you will find the **Math Post-Secondary Preparedness Package (MP³) Solutions**.

Each solution is one page, consisting of the original problem, a very clear solution, a note or two on the math, a common error, a practice problem or two with answers, and a URL visit if you need more background on that problem.

Do this package. Now. When you begin school, you will, mathematically, hit the ground running!

Don't leave this till school starts. Once you are on campus, you will have between 18 and 30 hours of classes and labs, tons of homework and assignments, tests almost immediately, plus a normal life to lead. You won't have time for the Preparedness Package. Do this package now while your distractions are minimal.

Believe me, it will help you be a university and/or college success. It will make your life infinitely better. It will make your professors' lives better.

A final note: if you find an error or there is a topic we didn't cover that you believe needs to be in the Math Post-Secondary Preparedness Package, then please let us know. ASAP!

Good luck and best wishes,
Professor Jack Weiner
Department of Mathematics and Statistics
University of Guelph
jweiner@uoguelph.ca

Professor Hosh Pesotan
Department of Mathematics and Statistics
University of Guelph
hpesotan@uoguelph.ca

Sylvia Nguyen and George Hutchinson
Mathematics Majors, Year 3
Department of Mathematics and Statistics
University of Guelph

Part I: Brushing up on Numerical Skills

1) Adding and Subtracting Fractions

Problem: Evaluate **without a calculator!**

(i) $\dfrac{2}{3} + 2\dfrac{5}{6}$ (ii) $3\dfrac{2}{5} - 2\dfrac{3}{4}$ (iii) $\dfrac{1}{6a} + \dfrac{2}{3a} - \dfrac{5}{12a}$

2) Multiplying and Dividing Fractions

Problem: Evaluate without a calculator!

(i) $\dfrac{5}{3} \times \dfrac{12}{25}$ (ii) $1\dfrac{2}{5} \times \dfrac{3}{4}$ (iii) $3 \times \dfrac{4}{5}$ (iv) $\dfrac{\left(\dfrac{5}{3}\right)}{\left(\dfrac{25}{12}\right)}$ (v) $\dfrac{\left(\dfrac{4}{3}\right)}{5}$ (vi) $\dfrac{4}{\left(\dfrac{3}{5}\right)}$

3) Working with Decimals

Problem: Evaluate without a calculator!

(i) $1.02 + .023$ (ii) $1.02 - 2.57$ (iii) 1.2×0.5 (iv) $\dfrac{4.291}{3}$

(v) $\dfrac{4.291}{0.3}$ (vi) $\dfrac{0.004291}{0.03}$

4) Roots and Radicals

Problems: 1) Write as simplified mixed radicals:

(i) $\sqrt{40}$ (ii) $2\sqrt{27}$ (iii) $\sqrt{x^4 y^7}$ (iv) $\sqrt[3]{x^4 y^7}$

2) Write as entire radicals: (i) $3\sqrt{2}$ (ii) $\dfrac{4}{9}$ (iii) $xy^4 \sqrt{xy}$ (iv) $xy^4 \sqrt[3]{xy}$

3) Evaluate: (i) $\sqrt{121}$ (ii) $\left(\dfrac{27}{64}\right)^{2/3}$ (iii) $32^{1/5}$ (iv) $(-32)^{1/5}$ (v) $(-64)^{1/6}$

5) Absolute Value

Problems: 1) Evaluate: (i) $|10|$ (ii) $|-10|$ (iii) $|0|$

2) Write $|x|$ without absolute value bars if (i) $x > 0$ (ii) $x < 0$.

3) Draw the graph of $y = |x|$.

Part II: Lines and Slopes

1) Finding the Slope of a Line

Problem: Find the slope of the line joining (2,5) to (7,4).

2) Parallel and Perpendicular Slopes

Problem: Find the slope of the line

(i) parallel (ii) perpendicular

to a line l with slope 2.

3) Interpreting Slope

Problem:

(i) Match the slopes $0, \dfrac{1}{2}, 1, 2$

with the lines l_1, l_2, l_3, l_4.

(ii) Match the slopes $-\dfrac{1}{3}, -1, -3$

with the lines l_5, l_6, l_7.

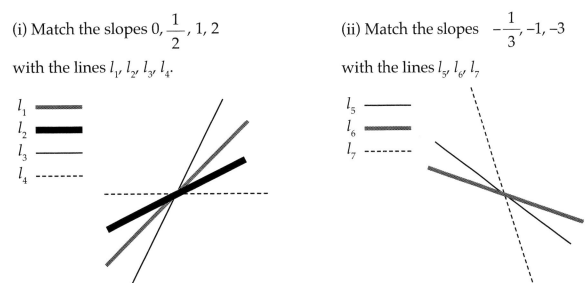

4) Finding the Slope and Intercepts From the Equation of a Line

Problem: Find the slope and x and y intercepts for each of the following lines:

(i) $y = -2x + 5$ (ii) $6x + 2y = 1$ (iii) $y = 5$ (iv) $x = 1$

5) Finding the Equation of a Line Given Two Points

Problem: Find the equation of the line joining (4,5) to (7,14). Draw the graph.

6) Finding the Equation of a Line Given the Slope and a Point

Problem: Find the equation of the line with slope $m = -2$ passing through the point $(-4,5)$. Draw the graph.

7) Graphing Linear Inequalities

Problem: Show by shading the region in the xy plane that satisfies
(i) $x + y \leq 3$ (ii) $2x - y > 4$.

Part III: Algebraic Skills

1) Adding and Subtracting Like Terms

Problem: Simplify: (i) $4st^2 + 2t^2s$

(ii) $(x^2 - 3xy + 7x - 1) + (2x^2 - xy - 3y - 4)$

(iii) $3(x + z) + 7(x + z) - y(x + z)$

2) Mulitplying Binomials

Problem: Expand: (i) $(3x + 1)(2x - 5)$ (ii) $(2a + 3b)^2$

3) Multiplying Binomials and Trinomials

Problem: Expand: (i) $(x^2 + 3x + 1)(2x - 5)$ (ii) $(a + b + c)^2$

4) Expanding $(a \pm b)^3$

Problem: Expand: (i) $(a + b)^3$ (ii) $(a - b)^3$

5) Factoring Easy Trinomials

Problem: Factor: (i) $x^2 + 5x + 4$ (ii) $x^2 + 3x - 4$ (iii) $6x^2 + 17x + 5$ (iv) $6x^2 - 13x - 5$

6) Factoring Less Easy Trinomials Using the Quadratic Formula

Problem: Factor: (i) $x^2 + 3x + 1$ (ii) $6x^2 - 5x - 2$

7) Factoring difference of squares: $a^2 - b^2 = (a - b)(a + b)$

Problem: 1) Factor: (i) $x^2 - 9$ (ii) $x^4 - (y+1)^2$

2) Rationalize the denominator: $\dfrac{1}{\sqrt{x} - 4}$

8) **Factoring difference of cubes:** $a^3 - b^3 = (a - b)(a^2 + ab + b^2)$

Problem: (i) Factor: $8x^3 - 27$

(ii) Rationalize the denominator: $\dfrac{1}{\sqrt[3]{x} - 2}$

9) **Factoring sum of cubes:** $a^3 + b^3 = (a + b)(a^2 - ab + b^2)$

Problem: (i) Factor: $8x^3 + 27$

(ii) Rationalize the denominator: $\dfrac{1}{\sqrt[3]{x} + 2}$

10) **Factoring** $a^n - b^n = (a - b)(a^{n-1} + a^{n-2}b + a^{n-3}b^2 + a^{n-4}b^3 + \ldots + ab^{n-2} + b^{n-1})$

Problem: Factor: $x^5 - y^5$

11) **Factoring** $a^n + b^n = (a + b)(a^{n-1} - a^{n-2}b + a^{n-3}b^2 - a^{n-4}b^3 + \ldots - ab^{n-2} + b^{n-1})$
and n MUST BE ODD!

Problem: Factor: $x^5 + y^5$

12) **The Factor Theorem: Part 1**

Problem: Factor the expression $x^3 - 4x^2 + x + 6$.

13) **The Factor Theorem: Part 2**

Problem: Find all the rational roots of $2x^3 - 5x^2 - 4x + 3$.

14) **Polynomial Division**

Problem: Divide $x^3 - 5x^2 + x + 6$ by $x - 4$; express your answer $\dfrac{x^3 - 5x^2 + x + 6}{x - 4}$ both in

the form $\dfrac{x^3 - 5x^2 + x + 6}{x - 4} = \text{Quotient} + \dfrac{\text{Remainder}}{\text{Divisor}}$

and $x^3 - 5x^2 + x + 6 = \text{Quotient} \times \text{Divisor} + \text{Remainder}$.

15) Solving for the Roots of a Polynomial

Problem: Solve: (i) $x^2 - 5x + 6 = 0$ (ii) $x^2 - 5x + 1 = 0$ (iii) $x^3 - 7x = -6$

16) Solving "Factored" Inequalities: Numerators Only

Problem: Solve: (i) $(x + 1)(x - 1)(x - 2) \geq 0$ (ii) $(x+1)^2 x(x-1)^3 > 0$

17) Solving "Factored" Rational Inequalities:

Problem: Solve: (i) $\dfrac{(x + 1)(x - 2)}{x - 1} \geq 0$ (ii) $\dfrac{(x + 1)^2(x - 1)^3}{x} > 0$

18) Completing the Square

Problem: Complete the square: (i) $x^2 - 8x + 25$ (ii) $3x^2 + 36x - 17$ (iii) $-2x^2 + 3x + 1$

19) Adding and Subtracting Rational Expressions

Problem: By getting a common denominator, simplify the following expressions:

(i) $\dfrac{1}{3x} - \dfrac{1}{2y} + \dfrac{1}{6z}$ (ii) $\dfrac{2x + 1}{x - 1} - \dfrac{x + 1}{x + 2} - \dfrac{5x + 4}{x^2 + x - 2}$

20) Multiplying and Dividing Rational Expressions

Problem: Simplify the following rational expressions:

(i) $\dfrac{(x^2 - 16)}{(x - 4)^3} \times \dfrac{(x^2 - 4x)^2}{x^3 + 64}$ (ii) $\dfrac{x^2 + 5xy + 4y^2}{x^2 + 4xy + 4y^2} \div \dfrac{x^2 + xy}{x^2 + 2xy}$

Part IV: Geometry

1) Pythagorean Theorem

Problem: In the following diagrams find the value of the unknowns:

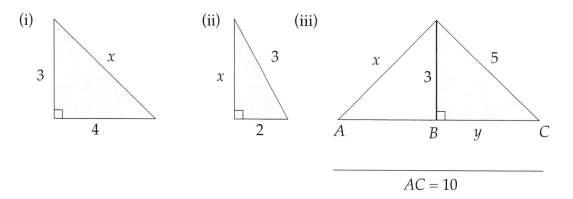

(i)   (ii)   (iii)

$AC = 10$

2) Angles in a Triangle

Problem: Find the values of angles x and y (in degree measure) from the following diagrams:

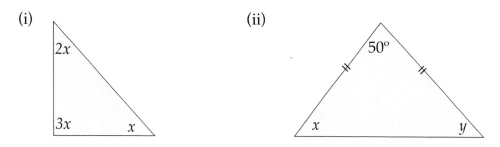

(i)   (ii)

3) The Parallel Line Theorem

Problem: Find the values in degrees of x and y in the following diagrams:

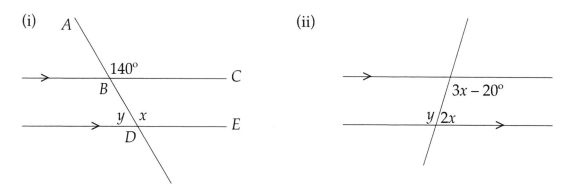

(i)   (ii)

4) Congruent Triangles

Problem: Triangles I and II are congruent. Name the congruent triangles so that the vertices "correspond" and determine the values x, y, u, and v.

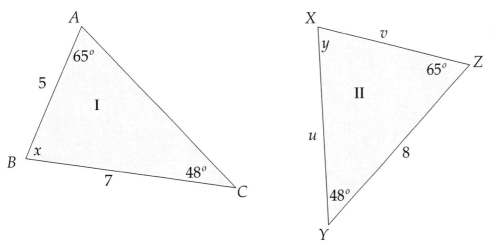

5) Similar Triangles

Problem: Triangles I and II are similar. Name the similar triangles so that the vertices "correspond" and determine the values x, y, and z.

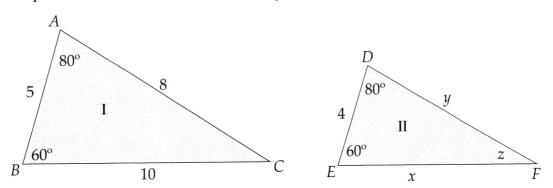

6) Area of a Triangle

Problem: Find the area of $\triangle ABC$ in each of the following:

(i)

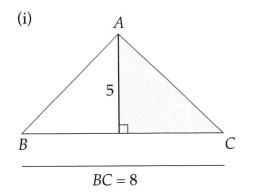

$$BC = 8$$

(ii)

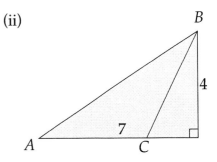

7) Area and circumference of a circle

Problem: (i) Find the circumference and area of a circle of radius 8 cm.
(ii) Find the circumference and area of a circle of diameter 8 m.
(iii) If the area of a circle is 16π, find the radius.

8) Arc Length and Area of a Sector of a Circle

Problem: (i) State the arc length s and the area A of the sector of this circle. Assume θ is in **radian** measure.

(ii) In the circle below, find the arc length s and area A

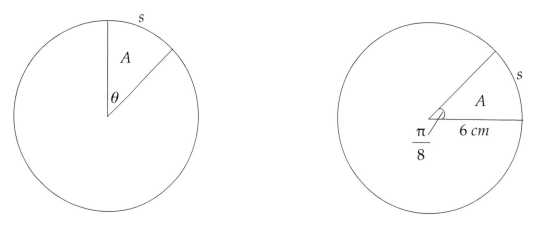

9) Volume of a Sphere, Box, Cone, Cylinder

Problem: Find the volume V of
(i) a sphere of radius $r = 3$ cm
(ii) a rectangular box with length $l = 8$ cm, width $w = 5$ cm, and height $h = 50$ cm
(iii) a right circular cone with height $h = 3$ cm and base radius $r = 0.04$ m
(iv) a circular cylinder with height $h = 0.2$ m and radius $r = 4$ cm.

10) Angles of a Polygon

Problem: i) Find the sum of the interior angles of a pentagon.

(ii) If all the interior angles of a pentagon are equal how much is each interior angle?

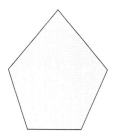

Part V: Basic Graphs

1) Graphing $y = x^n$, $n \in \mathbf{N}$

Problem: (i) Graph the curves $y = x^2$ and $y = x^4$ on the same set of axes.

(ii) Graph the curves $y = x^3$ and $y = x^5$ on the same set of axes.

2) Graphing $y = x^{m/n}$, $m, n \in \mathbf{N}$, n Odd, and m/n is a Reduced Fraction

Problem: (i) Graph the curves $y = x^{1/3}$ and $y = x^{2/3}$ on the same set of axes.

(ii) Graph the curves $y = x^{4/3}$ and $y = x^{5/3}$ on the same set of axes.

3) Graphing $y = x^{m/n}$, $m, n \in \mathbf{N}$, n Even, and m/n is a Reduced Fraction

Problem: Graph the curves $y = x^{1/2}$ and $y = x^{3/2}$ on the same set of axes.

4) Graphing $y = x^{-n} = \dfrac{1}{x^n}$, $n \in \mathbf{N}$

Problem: Graph the curves $y = x^{-1} = \dfrac{1}{x}$ and $y = x^{-2} = \dfrac{1}{x^2}$ on the same set of axes.

5) Graphing $y = x^{-1/n} = \dfrac{1}{x^{1/n}}$, $n \in \mathbf{N}$

Problem: Graph the curves $y = \dfrac{1}{x^{1/2}}$ and $y = \dfrac{1}{x^{1/3}}$

6) Transformations (New Graphs from a Given Graph)

Problem: Given $y = f(x) = x^2$, graph and describe each of the following relative to f:

(i) $y = f(x) + 2$　　(ii) $y = f(x) - 2$　　(iii) $y = f(x + 2)$　　(iv) $y = f(x - 2)$

(v) $y = f(2x)$　　(vi) $y = 2f(x)$

Part VI: Solving Equations and Inequalities

1) Solving Linear Equations in One Variable

Problem: Solve each of the following equations:

(i) $4x + 20 = 2 - 5x$ (ii) $8(x - 4) + x = 6(x - 5) - (1 - x)$ (iii) $\dfrac{x}{3} - \dfrac{2x}{5} + \dfrac{1}{30} = \dfrac{7}{10}$

2) Solving Linear Inequalities

Problem: Solve the following inequalities:

(i) $4x - 5 \leq 2x + 9$ (ii) $2x + 7 > 5x - 1$ (iii) $x - 4 < x + 6$

3) Solving Two Linear Equations in Two Variables

Problem: Solve for x and y: (E1 and E2 refer to equation 1 and equation 2.)

(i) E1: $x + 2y = -1$ (ii) E1: $x - 2y = 6$ (iii) E1: $x - 2y = 6$

   E2: $5x - 2y = 7$ E2: $3x - 6y = 18$ E2: $3x - 6y = 3$

4) Solving Quadratic Equations

Problem: Solve the following quadratic equations:

(i) $x^2 - 5x + 6 = 0$ (ii) $3x^2 - 7x + 2 = 0$

5) Solving Equations Involving Square Roots

Problem: Solve the following equations for x:

(i) $\sqrt{x - 2} = 5$ (ii) $\sqrt{4 - 3x} = x + 12$ (iii) $\sqrt{1 + 2x} - \sqrt{x} = 1$

Part VII: Graphing Second Order Relations

1) The Parabola

Problem: Graph the following parabolas and identify the vertex and the axis of symmetry: (i) $y = x^2$ (ii) $y = 2(x + 1)^2 - 3$

2) The Circle

Problem: Graph the following circles and identify the radius and the centre:
(i) $x^2 + y^2 = 4$ (ii) $(x - 1)^2 + (y + 2)^2 = 9$

3) The Ellipse

Problem: Graph the following ellipses and identify the centre and the major and minor axes:

(i) $\dfrac{x^2}{4} + \dfrac{y^2}{9} = 1$ (ii) $\dfrac{(x - 1)^2}{4} + \dfrac{(y + 2)^2}{9} = 1$

4) The Hyperbola

Problem: Graph the following hyperbolas and identify the centre and intercepts:

(i) $\dfrac{x^2}{4} - \dfrac{y^2}{9} = 1$ (ii) $\dfrac{y^2}{9} - \dfrac{x^2}{4} = 1$

Part VIII: Trigonometry

1) Angles in Standard Position: Degree Measure

Problem: (i) Draw in standard position the following angles:
(a) 30° (b) 225° (c) −80° (d) −190° (e) 390°

2) The Meaning of π

Problem: In the circle below with radius r, you can "fit" three and a "little portion more" of a radius around half the circle.

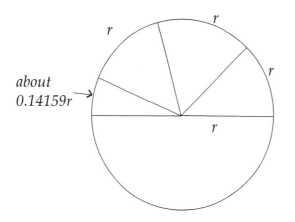

We give a name to this number of radii. (i) The name is _____.

(ii) So the length of a half circle is given by _____.

(iii) This is why the circumference is given by _____.

3) Angles in Standard Position: Radian Measure

Problems: 1) Draw in standard position the following angles:

(i) $\dfrac{\pi}{6}$ (ii) $\dfrac{5\pi}{4}$ (iii) $-\dfrac{4\pi}{9}$

2) Give all angles, in radians, which are "co-terminal" with π/6 radians.

4) Degrees to Radians

Problem: Express each of the following in radian measure:
(i) 25° (ii) −150° (iii) 1060° (Remember 180 degrees = π radians.)

5) Radians to Degrees

Problem: Express each of the following radian measures in degrees:

(i) $\dfrac{4\pi}{9}$ (ii) $\dfrac{2}{5}$ (iii) $-\dfrac{7\pi}{6}$ (iv) $-\dfrac{4\pi}{3}$ (Remember π radians = 180 degrees)

6) Relating Angles in Standard Position in Quadrants One and Two

(Related angles: angles whose trig ratios have the same magnitude but differ in sign according to the CAST RULE.)

Problem: (i) Below, the second quadrant angle 170° is drawn in standard position. Find and illustrate the related first quadrant angle, using the interval (0,90°).

(ii) Below, the first quadrant angle $\dfrac{\pi}{6}$ is drawn in standard position. Find and illustrate the related second quadrant angle, using the interval $\left(\dfrac{\pi}{2}, \pi\right)$.

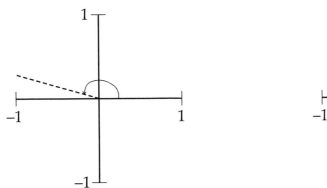 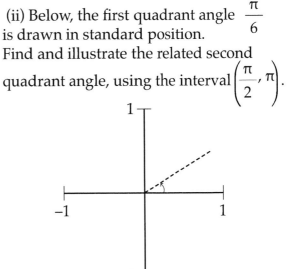

7) Relating Angles in Standard Position in Quadrants One and Three

(Related angles: angles whose trig ratios have the same magnitude but differ in sign according to the CAST RULE.)

Problem: (i) Below, the third quadrant angle 190° is drawn in standard position. Find and illustrate its related first quadrant angle, using the interval (0,90°).

(ii) Below, the first quadrant angle $\dfrac{\pi}{6}$ is drawn in standard position. Find and illustrate the related second quadrant angle, using the interval $\left(\pi, \dfrac{3\pi}{2}\right)$.

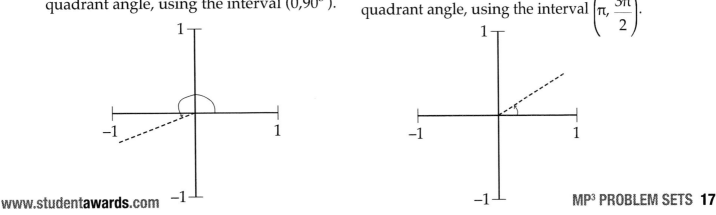

8) Relating Angles in Standard Position in Quadrants One and Four

(Related angles: angles whose trig ratios have the same magnitude but differ in sign according to the CAST RULE.)

Problem: (i) Below, the fourth quadrant angle −10° is drawn in standard position. Find and illustrate its related first quadrant angle using the interval (0,90°).

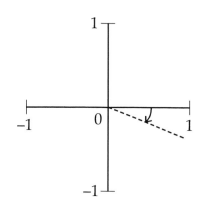

(ii) Below, the first quadrant angle $\dfrac{\pi}{6}$ is drawn in standard position. Find and illustrate its related fourth quadrant angle, using the interval $\left(-\dfrac{\pi}{2},0\right)$.

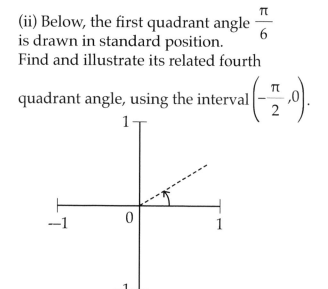

9) Relating an Angle in Standard Position to its "Relatives" in the other Quadrants

(Related angles: angles whose trig ratios have the same magnitude but differ in sign according to the CAST RULE.)

Quadrant	Angle
1	Degrees: $0° < \theta < 90°$ or Radians: $0 < \theta < \dfrac{\pi}{2}$
2	Degrees: $90° < \theta < 180°$ or Radians: $\dfrac{\pi}{2} < \theta < \pi$
3	Degrees: $180° < \theta < 270°$ or Radians: $\pi < \theta < \dfrac{3\pi}{2}$
4	Degrees: $-90° < \theta < 0°$ or Radians: $-\dfrac{\pi}{2} < \theta < 0$

Problem: State the standard position "relatives" of
(i) 50°   (ii) 170°   (iii) 250°   (iv) − 60° in each of the other quadrants.

10) Trigonometric Ratios in Right Triangles: SOHCAHTOA

Problem:

From the triangle, identify all six trigonometric ratios for θ

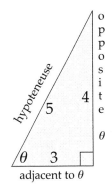

11) Trigonometric Ratios Using the Circle: Part I

Problem: Let θ be an angle which is not between $0°$ and $90°$. By drawing the angle in standard position and letting it puncture the unit circle $x^2 + y^2 = 1$ at a point (x, y), find the sin, cos, and tan of θ.

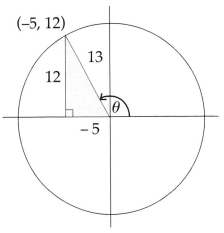

12) Trigonometric Ratios Using the Circle: Part II

Problem: In the diagram, θ, where $90° < \theta < 180°$, is a second quadrant angle in standard position whose terminal side punctures the circle (centered at the origin, with radius 13) at the point P($-5,12$). State the six trigonometric ratios of θ.

13) Trigonometric Ratios for the 45°, 45°, 90° Triangle

Problem: Find the values of all the six trigonometric ratios of $45° = \dfrac{\pi}{4}$

14) Trigonometric Ratios for the 30°, 60°, 90° Triangle

Problem: Find the values of all the six

trigonometric ratios of $60° = \dfrac{\pi}{3}$ and $30° = \dfrac{\pi}{6}$.

15) Trigonometric Ratios for the 0°, 90°, 180°, 270°; ie., Trig Ratios For Angles on the Axes

First please re-read "11) Trigonometric Ratios Using the Circle: Part I".

Problem: Find the six trig ratios for $0° = 0$ radians and $90° = \dfrac{\pi}{2}$ radians.

16) CAST RULE

Problems: 1) Given $\tan(\theta) = -4/3$, find the values of $\sin(\theta)$ and $\cos(\theta)$ if
(i) θ is a second quadrant angle.
(ii) θ is a fourth quadrant angle.
2) Why can't θ be a first or third quadrant angle?

17) Sine Law: Find an Angle

Problem: Find $\angle C$ and $\angle A$.

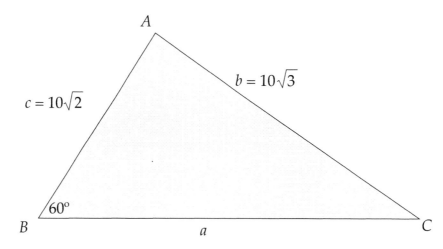

18) Sine Law: Find a Side

Problem: Find (i) the exact value of a and
(ii) c accurate to two decimals.

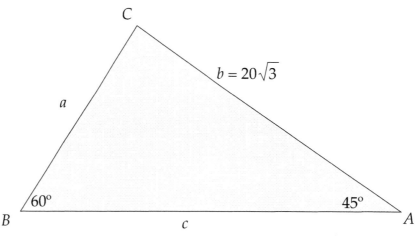

19) Cosine Law: Find an angle

Problem: In $\triangle ABC$, use the Cosine Law to find $\angle B$ to the nearest degree.

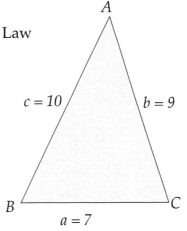

20) Cosine Law: Find a Side

Problem: In $\triangle ABC$, use the Cosine Law to find $c = AB$ correct to two decimal places.

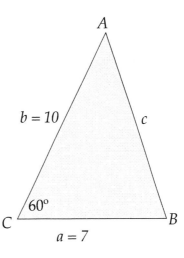

21) The Graphs of the Sin, Cos, and Tan Functions

Problem: Graph for $-2\pi \le \theta \le 2\pi$: (i) $y = \sin(\theta)$ (ii) $y = \cos(\theta)$ (iii) $y = \tan(\theta)$

22) Period of the Sin, Cos, and Tan Functions

Problems: State the period of each of the following:

1) (i) $y = \sin(x)$ (ii) $y = \sin(2x)$ (iii) $y = \sin\left(\dfrac{x}{2}\right)$

2) (i) $y = \cos(x)$ (ii) $y = \cos(3x)$ (iii) $y = \cos\left(\dfrac{x}{3}\right)$

3) (i) $y = \tan(x)$ (ii) $y = \tan(4x)$ (iii) $y = \tan\left(\dfrac{\pi}{2}x\right)$

23) The Graphs of the Csc, Sec, and Cot Functions

Problem: Graph for $-2\pi \le \theta \le 2\pi$: (i) $y = \csc(\theta)$ (ii) $y = \sec(\theta)$ (iii) $y = \cot(\theta)$

24) Trig Formulas That You Should Know

Problem: Complete the following formulas:

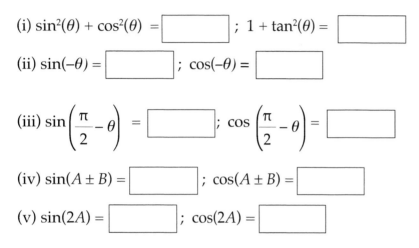

(i) $\sin^2(\theta) + \cos^2(\theta) \ =$ [ ] ; $1 + \tan^2(\theta) =$ [ ]

(ii) $\sin(-\theta) =$ [ ] ; $\cos(-\theta) =$ [ ]

(iii) $\sin\left(\dfrac{\pi}{2} - \theta\right) \ =$ [ ] ; $\cos\left(\dfrac{\pi}{2} - \theta\right) =$ [ ]

(iv) $\sin(A \pm B) =$ [ ] ; $\cos(A \pm B) =$ [ ]

(v) $\sin(2A) =$ [ ] ; $\cos(2A) =$ [ ]

Part IX: Exponents and Logarithms

1) Exponents

Problems:

1) Evaluate: (i) 2^3 (ii) $\left(\dfrac{3}{5}\right)^3$ (iii) 4^{-2} (iv) 10^0 (v) $\left(\dfrac{1}{0.01}\right)^{-3}$

2) Simplify: (i) $\dfrac{x^5 x^4}{x^7}$ (ii) w^{-1} (iii) $\dfrac{1}{w^{-3}}$ (iv) $\left(z^{2/3}\right)^{10}$ (v) $\left(\dfrac{a^7 b^3}{c^2}\right)^5$

2) Logarithms (Log means FIND THE EXPONENT!)

Problems: 1) Evaluate: (i) $\log_2 8$ (ii) $\log_2\left(\dfrac{1}{8}\right)$ (iii) $\log_3 1$ (iv) $\log_5 5$

(v) $\log 10000$ (vi) $\ln(e^7)$ *

2) Expand using log properties: $\ln\left(\dfrac{x^3 y^{1/2}}{z^4}\right)$

3) Change $\log_5 7$ to log with base 3, then with base 10, and finally with base e.

> * "e" and "ln" refer to the "natural logarithm". If you have not taken calculus, you may be totally unfamiliar with e. If so, treat it as a constant just as you would, for example, a.

3) Exponential Graphs

Problem: (i) Graph the exponential functions $y = 2^x$ and $y = 3^x$ on the same set of axes.

(ii) Graph the exponential functions $y = 2^{-x} = \dfrac{1}{2^x}$ and $y = 3^{-x} = \dfrac{1}{3^x}$ on the same set of axes.

4) Logarithmic Graphs

Problem: Graph the functions $y = \log_2(x)$ and $y = \log_3(x)$ on the same set of axes.

5) Exponents to Logarithms and Vice-Versa

Problems: 1) Change to a log equation: (i) $32 = 2^5$ (ii) $y = 10^x$ (iii) $y = 4x^k$ (Use base 10.)

2) Change to an exponential equation: (i) $\log_3 81 = 4$ (ii) $y = \log_5 x$

6) Using a Calculator to Evaluate Exponents and Logs

Problems: Use a calculator to give answers rounded to two decimal places.

1) (i) $2^{3.3}$ (ii) $5^{1/7}$ (iii) $(-10)^{1/3}$

2) (i) $\log(25)$ (ii) $\ln(25)$ *   (iii) $\log_2(25)$

---

* "e" and "ln" refer to the "natural logarithm". If you have not taken calculus, you may be totally unfamiliar with e. If so, treat it as a constant just as you would, for example, a.

ANSWERS
Additional practice
and web references

ANSWERS
Additional practice and web references

Part I: Brushing up on Numerical Skills

Part I: Brushing up on numerical skills

1) Adding and Subtracting Fractions

Problem: Evaluate **without a calculator!**

(i) $\dfrac{2}{3} + 2\dfrac{5}{6}$ (ii) $3\dfrac{2}{5} - 2\dfrac{3}{4}$ (iii) $\dfrac{1}{6a} + \dfrac{2}{3a} - \dfrac{5}{12a}$

Solution:

(i) $\dfrac{2}{3} + 2\dfrac{5}{6} = \dfrac{4}{6} + \dfrac{17}{6} = \dfrac{21}{6} = \dfrac{7}{2} \overset{\boxed{\text{optional}}}{=} 3\dfrac{1}{2}$

(ii) $3\dfrac{2}{5} - 2\dfrac{3}{4} = \dfrac{68}{20} - \dfrac{55}{20} = \dfrac{13}{20}$ or $3\dfrac{2}{5} - 2\dfrac{3}{4} = 1\dfrac{2}{5} - \dfrac{3}{4} = \dfrac{28}{20} - \dfrac{15}{20} = \dfrac{13}{20}$

(iii) $\dfrac{1}{6a} + \dfrac{2}{3a} - \dfrac{5}{12a} = \dfrac{2}{12a} + \dfrac{8}{12a} - \dfrac{5}{12a} = \dfrac{5}{12a}$

Note: Use the "lowest" common denominator.

Common error: Not using the lowest common denominator leads to larger numerators and more possibility of making mechanical errors.

Practice: (i) $\dfrac{3}{4} + 1\dfrac{1}{2}$ (ii) $3\dfrac{1}{7} - 1\dfrac{2}{3}$ (iii) $\dfrac{3}{5} + \dfrac{2}{4} - \dfrac{1}{30}$

Answer: (i) $\dfrac{9}{4} = 2\dfrac{1}{4}$ (ii) $\dfrac{31}{21} = 1\dfrac{10}{21}$ (iii) $\dfrac{16}{15}$

A Great Website for More Detail: http://www.math.com/

2) Multiplying and Dividing Fractions

Problem: Evaluate **without a calculator!**

(i) $\dfrac{5}{3} \times \dfrac{12}{25}$ (ii) $1\dfrac{2}{5} \times \dfrac{3}{4}$ (iii) $3 \times \dfrac{4}{5}$ (iv) $\dfrac{\left(\dfrac{5}{3}\right)}{\left(\dfrac{25}{12}\right)}$ (v) $\dfrac{\left(\dfrac{4}{3}\right)}{5}$ (vi) $\dfrac{4}{\left(\dfrac{3}{5}\right)}$

Solution:

(i) $\dfrac{\cancel{5}^{1}}{\cancel{3}_{1}} \times \dfrac{\cancel{12}^{4}}{\cancel{25}_{5}} = \dfrac{1 \times 4}{1 \times 5} = \dfrac{4}{5}$ (ii) $1\dfrac{2}{5} \times \dfrac{3}{4} = \dfrac{7}{5} \times \dfrac{3}{4} = \dfrac{21}{20} \overset{\boxed{\text{optional}}}{=} 1\dfrac{1}{20}$

(iii) $3 \times \dfrac{4}{5} = \dfrac{3}{1} \times \dfrac{4}{5} = \dfrac{12}{5} = 2\dfrac{2}{5}$ (iv) $\dfrac{\left(\dfrac{5}{3}\right)}{\left(\dfrac{25}{12}\right)} \overset{\boxed{\text{invert and multiply}}}{=} \dfrac{\cancel{5}^{1}}{\cancel{3}_{1}} \times \dfrac{\cancel{12}^{4}}{\cancel{25}_{5}} = \dfrac{1 \times 4}{1 \times 5} = \dfrac{4}{5}$

(v) $\dfrac{\left(\dfrac{4}{3}\right)}{5} = \dfrac{\left(\dfrac{4}{3}\right)}{\left(\dfrac{5}{1}\right)} = \dfrac{4}{3} \times \dfrac{1}{5} = \dfrac{4}{15}$ (vi) $\dfrac{4}{\left(\dfrac{3}{5}\right)} = \dfrac{\left(\dfrac{4}{1}\right)}{\left(\dfrac{3}{5}\right)} = \dfrac{4}{1} \times \dfrac{5}{3} = \dfrac{20}{3}$

Note: Common denominators are irrelevant when multiplying and dividing.

Common error: $\dfrac{\left(\dfrac{a}{b}\right)}{c} = \dfrac{a}{b} \times \dfrac{c}{1}$

Practice: (i) $\dfrac{3}{4} \times \dfrac{8}{15}$ (ii) $\dfrac{\left(\dfrac{2}{7}\right)}{\left(\dfrac{12}{21}\right)}$ (iii) $\dfrac{\left(\dfrac{22}{3}\right)}{11}$ (iv) $\dfrac{22}{\left(\dfrac{3}{11}\right)}$

Answer: (i) $\dfrac{2}{5}$ (ii) $\dfrac{1}{2}$ (iii) $\dfrac{2}{3}$ (iv) $\dfrac{242}{3}$

A Great Website for More Detail: http://www.math.com/

3) Working with Decimals.

Problem: Evaluate **without a calculator!**

(i) $1.02 + .023$ (ii) $1.02 - 2.57$ (iii) $1.2 \times .05$ (iv) $\dfrac{4.291}{3}$

(v) $\dfrac{4.291}{0.3}$ (vi) $\dfrac{0.004291}{0.03}$

Solution:

(i) $1.02 + .023 \quad \boxed{\text{Align the decimals}} \quad = \quad 1.043$ (ii) $1.02 - 2.57 = -(2.57 - 1.02) = -1.55$

$\boxed{\text{Work out } 12 \times 5 = 60. \text{ Now start on the right and count THREE decimal places to the left.}}$

(iii) $1.2 \times .05 \quad = \quad 0.06$ (iv) $\dfrac{4.291}{3} = 1.430\overline{3} = 1.4303333\ldots$

$\boxed{\text{Multiply the bottom by 10 to move the decimal after the 3. Multiply the top by 10 so we don't change the value.}}$

(v) $\dfrac{4.291}{0.3} \quad = \quad \dfrac{4.291}{0.3} \times \dfrac{10}{10} = \dfrac{42.91}{3} = 14.30\overline{3}$

(vi) $\dfrac{0.004291}{0.03} = \dfrac{0.004291}{0.03} \times \dfrac{100}{100} = \dfrac{.4291}{3} = 0.1430\overline{3}$

Note: When dividing by a decimal, most of us were taught to move the decimal the same number of places to the right in questions such as (iv) and (v). What we are really doing is multiplying the numerator and denominator by the same power of 10.

Common error: Running to our calculators when there is a decimal in the denominator because division by a decimal is scary.

Practice: (i) $21.021 - 25.555$ (ii) 30×0.00005 (iii) $\dfrac{3.6}{.000018}$

Answer: (i) -4.534 (ii) 0.0015 (iii) $200\,000$

A Great Website for More Detail: http://www.math.com

4) Roots and Radicals

Problems: 1) Write as simplified mixed radicals:

(i) $\sqrt{40}$ (ii) $2\sqrt{27}$ (iii) $\sqrt{x^4y^7}$ (iv) $\sqrt[3]{x^4y^7}$

2) Write as entire radicals: (i) $3\sqrt{2}$ (ii) $\dfrac{4}{9}$ (iii) $xy^4\sqrt{xy}$ (iv) $xy^4\sqrt[3]{xy}$

3) Evaluate: (i) $\sqrt{121}$ (ii) $\left(\dfrac{27}{64}\right)^{2/3}$ (iii) $32^{1/5}$ (iv) $(-32)^{1/5}$ (v) $(-64)^{1/6}$

Solution:

1)(i) $2\sqrt{10}$ (ii) $6\sqrt{3}$ (iii) $x^2y^3\sqrt{y}$ (iv) $xy^2\sqrt[3]{xy}$

2)(i) $\sqrt{18}$ (ii) $\sqrt{\dfrac{16}{81}}$ (iii) $\sqrt{x^3y^9}$ (iv) $\sqrt[3]{x^4y^{13}}$

3)(i) 11 (ii) $\dfrac{9}{16}$ (iii) 2 (iv) -2 (v) does not exist (as a "real number")

Note: Radical signs are just special cases of the more common exponents signs. For example, $\sqrt{x^3} = x^{3/2}$

Common error: In real numbers, we can not have the square root or fourth root or sixth root, etc., of a negative number. We CAN have the third root or the fifth root or the seventh root, etc., of a negative number. By the way, many calculators give an error message when ask for the odd root of a negative. Boo to the calculator's programmer!

Practice: 1) Write as mixed radicals: (i) $\sqrt{1000}$ (ii) $2\sqrt{27}$ (iii) $\sqrt{x^{12}y^9}$ (iv) $\sqrt[5]{x^{15}y^7}$

2) Write as entire radicals: (i) $5\sqrt{5}$ (ii) $\dfrac{4}{9}$ (iii) $xy^5\sqrt{x^{1/3}y}$ (iv) $x^2y^4\sqrt[4]{x}$

3) Evaluate: (i) $\sqrt[3]{-27}$ (ii) $\left(\dfrac{36}{25}\right)^{3/2}$ (iii) $(-243)^{1/5}$ (iv) $(-81)^{3/4}$

Answers: 1)(i) $10\sqrt{10}$ (ii) $6\sqrt{3}$ (iii) $x^6y^4\sqrt{y}$ (iv) $x^3y\sqrt[5]{y^2}$

2)(i) $\sqrt{125}$ (ii) $\sqrt{\dfrac{16}{81}}$ (iii) $\sqrt{x^{7/3}\,y^{11}}$ (iv) $\sqrt[4]{x^9y^{16}}$

3)(i) -3 (ii) $\dfrac{216}{125}$ (iii) -3 (iv) does not exist

A Great Website for More Detail: Gov.pe.ca/educ/docs/curriculum/521Bunit1.pdf

5) Absolute Value

Problems: 1) Evaluate: (i) $|10|$ (ii) $|-10|$ (iii) $|0|$

2) Write $|x|$ without absolute value bars if (i) $x > 0$ (ii) $x < 0$

3) Draw the graph of $y = |x|$

Solution: 1) (i) 10   (ii) 10   (iii) 0   2) (i) $|x| = x$ (ii) $|x| = -x$

3)

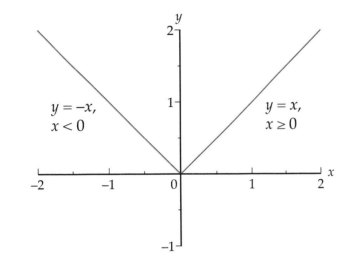

$y = -x,$
$x < 0$

$y = x,$
$x \geq 0$

Note: When, for $x < 0$, we write $|x| = -x$, remember there is a negative **inside the** x!

Common error: Assuming $|-x| = x$. This may or may not be true **depending on** x.
For example, if $x = 3$, then $|-x| = |-3| = 3 = x$. But if $x = -3$, then
$|-x| = |-(-3)| = |3| = 3 \neq x$; in fact, here $|-x| = -x$. It depends on x.

Practice: 1) Evaluate (i) $|-.001|$ (ii) $|.001|$ (iii) $|-0|$
2) Write $y = |x - 1|$ without using absolute value notation and graph the function.

Answers: 1) (i) .001 (ii) .001 (iii) 0

2) $y = \begin{cases} -(x-1), \text{ if } x < 1 \\ x - 1, \text{ if } x \geq 1 \end{cases}$

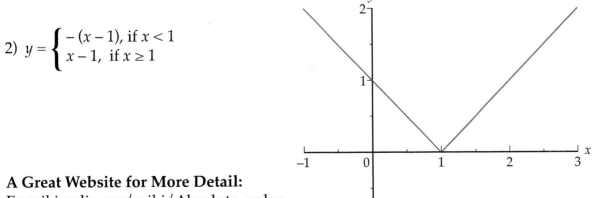

A Great Website for More Detail:
En.wikipedia.org / wiki / Absolute_value

ANSWERS
Additional practice
and web references

Part II: Lines and Slopes

Part II: Lines and slopes

1) Finding the slope of a line

Problem: Find the slope of the line joining (2,5) to (7,4).

Solution:

$$\text{slope} = \frac{\text{rise}}{\text{run}} = \frac{y_2 - y_1}{x_2 - x_1} = \frac{4-5}{7-2} = -\frac{1}{5}$$

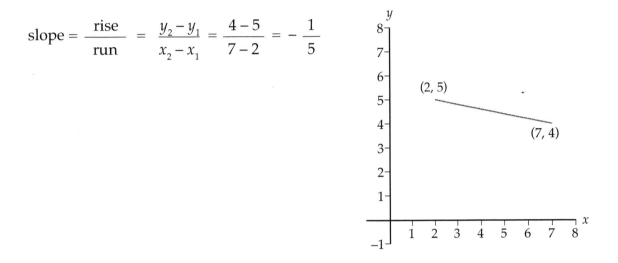

Note: It doesn't matter which point you use as (x_1, y_1).

Common error: $\text{slope} = \frac{\text{rise}}{\text{run}} = \frac{y_2 - y_1}{x_1 - x_2} = \frac{4-5}{2-7} = \frac{1}{5}$

Practice: Find the slope of the line joining $(-2, 8)$ to $(4, -5)$.

Answer: Slope $= -\dfrac{13}{6}$

A Great Website for More Detail: purplemath.com/modules/strtlneq.htm

2) Parallel and Perpendicular Slopes

Problem: Find the slope of the line

(i) parallel (ii) perpendicular

to a line l with slope 2.

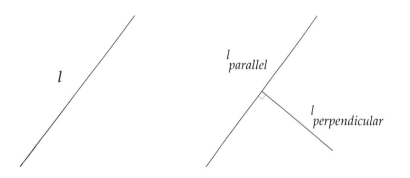

Solution: (i) Parallel slope $= 2$ (ii) Perpendicular slope $= -\dfrac{1}{2}$

Note: A line with slope 0 is horizontal. In this case, the perpendicular line, which is vertical, has undefined (or infinite) slope. Apart from the case of horizontal/vertical lines, the slope of a line perpendicular to a line with slope m satisfies

$m_{perpendicular} = -\dfrac{1}{m}$, that is, these slopes are "negative reciprocals".

Common error: Perpendicular slope $= \dfrac{1}{2}$

Practice: Find the slope of a line (i) parallel (ii) perpendicular to a line with slope $-\dfrac{3}{2}$.

Answer: (i) $-\dfrac{3}{2}$ (ii) $\dfrac{2}{3}$

A Great Website for More Detail: purplemath.com/modules/strtlneq.htm

3) Interpreting Slope

Problem:

(i) Match the slopes $0, \dfrac{1}{2}, 1, 2$

with the lines l_1, l_2, l_3, l_4.

(ii) Match the slopes $-\dfrac{1}{3}, -1, -3$

with the lines l_5, l_6, l_7

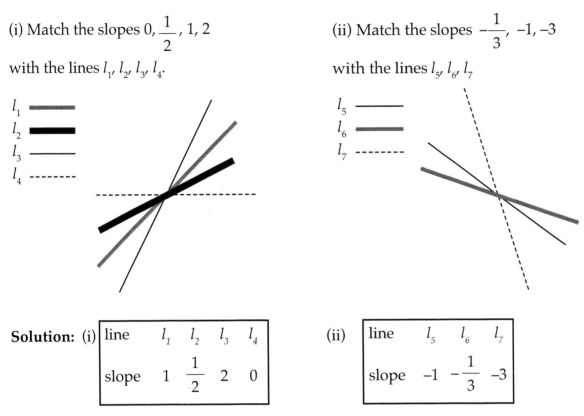

Solution: (i)

line	l_1	l_2	l_3	l_4
slope	1	$\dfrac{1}{2}$	2	0

(ii)

line	l_5	l_6	l_7
slope	-1	$-\dfrac{1}{3}$	-3

Note: For positive slope, as x increases, y increases. For negative slope, as x increases, y decreases. As you walk **to the right**, with positive slope you are walking uphill. With negative slope, you are walking downhill.

Common error: Misidentifying slope because the axes are not scaled one to one.

Practice: Match the slopes
0 and undefined (or ∞)
with the lines l_1 and l_2.

l_1 ———
l_2 ▬ ▬

Answer: l_1 has undefined (or infinite) slope and l_2 has 0 slope.

A Great Website for More Detail: algebra-online.com/slope-lines-1.htm

4) Finding the Slope and Intercepts From the Equation of a Line

Problem: Find the slope and x and y intercepts for each of the following lines:

(i) $y = -2x + 5$ (ii) $6x + 2y = 1$ (iii) $y = 5$ (iv) $x = 1$

Solution: In the equation $y = mx + b$, m is the slope and b is the y intercept.

(i) $y = -2x + 5$: $m = -2$ and $b = 5$. To find the x intercept, set $-2x + 5 = 0 \Rightarrow x = \dfrac{5}{2}$.

(ii) $6x + 2y = 1$: x intercept $\overset{\boxed{\text{Set } y = 0}}{\Rightarrow}$ $6x = 1 \Rightarrow x = \dfrac{1}{6}$

y intercept $\overset{\boxed{\text{Set } x = 0!}}{\Rightarrow}$ $2y = 1 \Rightarrow y = \dfrac{1}{2}$

slope $\overset{\boxed{\text{Rewrite the equation in the form } y = mx + b.}}{\Rightarrow}$ $y = -3x + \dfrac{1}{2} \Rightarrow m = -3$

(iii) $y = 5 = 0x + 5$ This is a horizontal line, that is, a line parallel to the axis. The slope m is 0, the y intercept is 5, and there is no x intercept.

(iv) $x = 1$ This is a vertical line, that is, a line parallel to the y axis. The slope is infinite (or undefined). The x intercept is 1 and there is no y intercept.

Note: The y intercept corresponds to the point $(0, y)$ on the line and the x intercept corresponds to $(x, 0)$.

Common error: Confusing the y intercept b with the point $(0, b)$.

Practice: Find the slope and x and y intercepts of the line $y = -2x + 6$.

Answer: $m = -2$; x intercept $= 3$; y intercept $= 6$

A Great Website for More Detail:

Math.com/school/subject2/lessons/S2U4L2GL.html

5) Finding the Equation of a Line Given Two Points

Problem: Find the equation of the line joining (4,5) to (7,14). Draw the graph.

Solution: The easiest formula for finding the equation of a line with slope m and point (x_1, y_1) is $y - y_1 = m(x - x_1)$

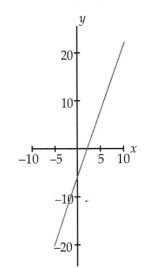

$$m = \frac{y_2 - y_1}{x_2 - x_1} = \frac{14 - 5}{7 - 4} = 3 \quad \text{Use } (4, 5) \text{ as } (x_1, y_1).$$

$$\therefore \ y - 5 = 3(x - 4) \text{ and so } y = 3x - 7$$

Note: It doesn't matter which point you use for (x_1, y_1).
If you are given the slope, you only need one point.

Common error: $y - 5 = 3(x-4)$ and so $y = 3x - 17$

Practice: Find the equation of the line joining $(-2, 1)$ to $(4, -5)$. Draw the graph.

Answer: $y = -x - 1$

A Great Website for More Detail: Math.com/school/subject2/lessons/S2U4L2DP.html

6) Finding the Equation of a Line Given the Slope and a Point

Problem: Find the equation of the line with slope $m = -2$ passing through the point $(-4, 5)$. Draw the graph.

Solution: The easiest formula for finding the equation of a line with slope m and point (x_1, y_1) is $y - y_1 = m(x - x_1)$

$m = -2$ and $(x_1, y_1) = (-4, 5)$

$\therefore y - 5 = -2(x + 4)$ and so $y = -2x - 3$

Note: Sometimes, you are given a "disguised" point. For example, if we were given x intercept $\dfrac{13}{2}$, we would use the point $\left(\dfrac{13}{2}, 0\right)$.

Common error: $y - 5 = -2(x - 4)$ and so $y = 3x + 3$

Practice: Find the equation of the line with slope $m = 5$ and with x intercept -4. Draw the graph.

Answer: $y = 5x + 20$

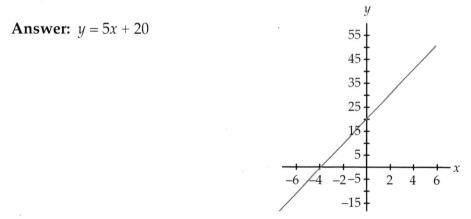

A Great Website for More Detail: Purplemath.com/modules/strtlneq.htm

7) Graphing Linear Inequalities

Problem: Show by shading the region in the xy plane that satisfies
(i) $x + y \leq 3$ (ii) $2x - y > 4$.

Solution: (i) Draw $x + y = 3$.
Test (0,0) in this equation
Left Side = 0; Right Side =3;
$0 \leq 3$ is **true**. So the shaded
region is on the same side of
the line as (0,0).

(ii) Draw $2x - y = 4$.
Test (0,0) in this equation
Left Side = 0; Right Side = 4;
$0 > 4$ is **false**. So the shaded region is on the **opposite** side of the
line as (0,0).

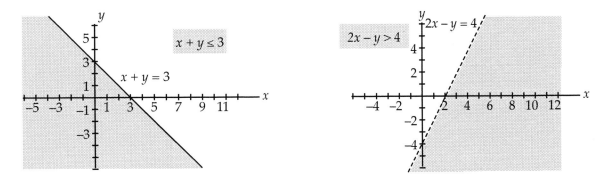

Note: For linear inequalities, draw the line first, dotted if the sign is < or >, solid if the sign is ≤ or ≥. Then test a point on one side of the line. If the point makes the inequality true, shade that portion of the plane. If not, shade the portion on the other side of the line.

Common error: Testing a point which is on the line. This won't answer the inequality question. You won't get LS < RS or LS > RS; you will get LS = RS, which is no help!

Practice: Show by shading the region in the xy plane that satisfies $x + 2y > 1$.

Answer:

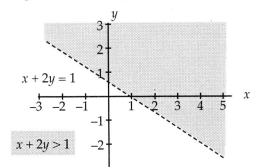

A Great Website for More Detail: Purplemath.com/modules/syslneq.htm

ANSWERS
Additional practice and web references

Part III: Algebraic Skills

Part III: Algebraic Skills

1) Adding and Subtracting Like Terms

Problem: Simplify: (i) $4st^2 + 2t^2s$

(ii) $(x^2 - 3xy + 7x - 1) + (2x^2 - xy - 3y - 4)$

(iii) $3(x + z) + 7(x + z) - y(x + z)$

Solution:

(i) $6st^2$

(ii) $(x^2 - 3xy + 7x - 1) + (2x^2 - xy - 3y - 4) = 3x^2 - 4xy + 7x - 3y - 5$

(iii) $3(x + z) + 7(x + z) - y(x + z) = (10 - y)(x + z)$

Notes: Terms are factors "glued" together with multiplication and division and are separated by addition and subtraction. In (iii), $(x + z)$ is a factor of each term. Don't be confused by the fact that it is composed of two terms inside the brackets.

Common error: In (i), **suppose the question were** $4st^2 + 2ts^2$. In this case, since $4st^2$ and $2ts^2$ are not like terms we cannot simplify the expression any further.

Practice: Simplify: $(4w + 3wx - 2) - (2w - 3wx + 1)$

Answer: $2w + 6wx - 3$

A Great Website for More Detail: cstl.syr.edu/fipse/Algebra/Unit2/plusminu.htm

2) Mulitplying Binomials

Problem: Expand: (i) $(3x + 1)(2x - 5)$ (ii) $(2a + 3b)^2$

Solution:

(i) $(3x + 1)(2x - 5) = 6x^2 - 15x + 2x - 5 = 6x^2 - 13x - 5$

(ii) $(2a + 3b)^2 = (2a + 3b)(2a + 3b) = 4a^2 + 12ab + 9b^2$

Note: Each separate term in the first bracket is multiplied with each term in the second. Watch the signs!

Common error: In (ii), $(2a + 3b)^2 = 4a^2 + 9b^2$

Practice: Expand: (i) $(a - 2b)(a + 2b)$ (ii) $(5w - 2)^2$

Answer: (i) $a^2 - 4b^2$ ("difference of squares") (ii) $25w^2 - 20w + 4$

A Great Website for More Detail:

regentsprep.org/regents/math/polymult/Smul_bin.htm

3) Multiplying Binomials and Trinomials

Problem: Expand: (i) $(x^2 + 3x + 1)(2x - 5)$ (ii) $(a + b + c)^2$

Solution: (i) $(x^2 + 3x + 1)(2x - 5) = 2x^3 - 5x^2 + 6x^2 - 15x + 2x - 5 = 2x^3 + x^2 - 13x - 5$

(ii) $(a + b + c)^2 = (a + b + c)(a + b + c) = a^2 + b^2 + c^2 + 2ab + 2ac + 2bc$

Note: The number of terms when you expand is the product of the number of terms in each bracket, that is, until you simplify. So there are $(3)(2) = 6$ terms in (i) and $(3)(3) = 9$ terms in (ii) before simplification.

Common error: In (ii), $(a + b + c)^2 = a^2 + b^2 + c^2$

Practice: Expand: (i) $(a + 2b - c)(a + 2b + c)$ (ii) $(a - b - c)^2$

Answer: (i) $a^2 + 4ab + 4b^2 - c^2$ (ii) $a^2 + b^2 + c^2 - 2ab - 2ac + 2bc$

A Great Website for More Detail:
k12connect.ca/~steve_soames/Lessons%20&%20Formulas/EXPRESSION/MultBin oTrino.htm

4) Expanding $(a \pm b)^3$

Problem: Expand: (i) $(a + b)^3$ (ii) $(a - b)^3$

Solution:

(i) $(a + b)^3 = (a + b)^2 (a + b) = (a^2 + 2ab + b^2)(a + b)$
$= a^3 + a^2b + 2a^2b + 2ab^2 + b^2a + b^3 = a^3 + 3a^2b + 3ab^2 + b^3$

(ii) $(a - b)^3 = (a - b)^2 (a - b) = (a^2 - 2ab + b^2)(a - b)$
$= a^3 - a^2b - 2a^2b + 2ab^2 + b^2a - b^3 = a^3 - 3a^2b + 3ab^2 - b^3$

Note: ab^2 and b^2a are "like" terms which is why $2ab^2 + b^2a = 3a^2b$.

Common error: In (i) $(a + b)^3 = a^3 + b^3$

Practice: Expand $(2x - 3y)^3$

Answer: $8x^3 - 36x^2y + 54xy^2 - 27y^3$

A Great Website for More Detail: purplemath.com/modules/binomial.htm

5) Factoring Easy Trinomials

Problem: Factor: (i) $x^2 + 5x + 4$ (ii) $x^2 + 3x - 4$ (iii) $6x^2 + 17x + 5$ (iv) $6x^2 - 13x - 5$

Solution:

(i) $x^2 + 5x + 4 = (x + 4)(x + 1)$

(ii) $x^2 + 3x - 4 = (x + 4)(x - 1)$

(iii) $6x^2 + 17x + 5 = (3x + 1)(2x + 5)$

(iv) $6x^2 - 13x - 5 = (3x + 1)(2x - 5)$

Note: $(ax + b)(cx + d) = acx^2 + (ad + bc)x + bd$

We need to find a and c for the x^2 coefficient and b and d for the constant so that $ad + bc$ is the x coefficient. We do this largely by trial and error. When stuck, there is a sure method. See the next question.

Common error: In (ii) $x^2 + 3x - 4 = (x - 4)(x + 1)$

Practice: Factor: (i) $x^2 - 2x - 15$ (ii) $9x^2 + 12x + 4$

Answer: (i) $(x - 5)(x + 3)$ (ii) $(3x + 2)^2$

A Great Website for More Detail: themathpage.com/alg/factoring-trinomials.htm

6) Factoring Less Easy Trinomials Using the Quadratic Formula

Problem: Factor: (i) $x^2 + 3x + 1$ (ii) $6x^2 - 5x - 2$

Hint: EVERY quadratic expression of the form $ax^2 + bx + c$ can be factored as

$a(x - r_1)(x - r_2)$, where $r_1 = \dfrac{-b + \sqrt{b^2 - 4ac}}{2a}$ and $r_2 = \dfrac{-b - \sqrt{b^2 - 4ac}}{2a}$.

Solution:

(i) If we solve $x^2 + 3x + 1 = 0$ using the quadratic formula with $a = 1$, $b = 3$ and $c = 1$, we find roots

$$r_1 = \frac{-3 + \sqrt{9 - 4(1)(1)}}{2} = \frac{-3 + \sqrt{5}}{2} \text{ and } r_2 = \frac{-3 - \sqrt{9 - 4(1)(1)}}{2} = \frac{-3 - \sqrt{5}}{2}$$

$$\therefore x^2 + 3x + 1 = (x - r_1)(x - r_2) = \left(x - \frac{-3 + \sqrt{5}}{2}\right)\left(x - \frac{-3 - \sqrt{5}}{2}\right) = \left(x - \frac{-3 + \sqrt{5}}{2}\right)\left(x + \frac{3 + \sqrt{5}}{2}\right)$$

(ii) If we solve $6x^2 - 5x - 2 = 0$ using the quadratic formula with

$a = 6$, $b = -5$ and $c = -2$, we find roots

$$r_1 = \frac{5 + \sqrt{25 - 4(6)(-2)}}{12} = \frac{5 + \sqrt{73}}{12} \text{ and } r_2 = \frac{5 - \sqrt{25 - 4(6)(-2)}}{12} = \frac{5 - \sqrt{73}}{12}$$

$$\therefore 6x^2 - 5x - 2 = 6(x - r_1)(x - r_2) = 6\left(x - \frac{5 + \sqrt{73}}{12}\right)\left(x - \frac{5 - \sqrt{73}}{12}\right)$$

Note: Note that in (ii) we put the original x^2 coefficient, 6, "out front" in the answer. We really reduced the problem to factoring $6\left(x^2 - \dfrac{5}{6}x - \dfrac{1}{3}\right)$ but working wth integers

for a, b and c is easier than working with fractions.

Common error: In (ii) $6x^2 - 5x - 2 = \left(x - \dfrac{5 + \sqrt{73}}{12}\right)\left(x - \dfrac{5 - \sqrt{73}}{12}\right)$ that is, forgetting to multiply by the x^2 coefficient 6.

Practice: Factor: (i) $x^2 - 2x - 15$ (ii) $3x^2 + 9x + 2$

Answer: (i) $(x - 5)(x + 3)$ (ii) $3\left(x - \dfrac{-9 + \sqrt{57}}{6}\right)\left(x + \dfrac{9 + \sqrt{57}}{6}\right)$

A Great Website for More Detail: purplemath.com/modules/quadform.htm

7) Factoring difference of squares: $a^2 - b^2 = (a - b)(a + b)$

Problem: 1) Factor: (i) $x^2 - 9$ (ii) $x^4 - (y + 1)^2$

2) Rationalize the denominator: $\dfrac{1}{\sqrt{x} - 4}$

Solution:

1) (i) $x^2 - 9 = (x - 3)(x + 3)$

(ii) $x^4 - (y + 1)^2 \overset{\boxed{\begin{array}{l} a = x^2 \\ b = y + 1 \end{array}}}{=} (x^2 - (y + 1))(x^2 + (y + 1)) = (x^2 - y - 1)(x^2 + y + 1)$

2) $\dfrac{1}{\sqrt{x} - 4} \overset{\boxed{\begin{array}{c}\text{Use difference of squares in reverse!} \\ a = \sqrt{x} \quad b = 4 \end{array}}}{=} \dfrac{\sqrt{x} + 4}{(\sqrt{x} - 4)(\sqrt{x} + 4)} = \dfrac{\sqrt{x} + 4}{x - 16}$

Note: The order of the factors doesn't matter: $a^2 - b^2 = (a - b)(a + b) = (a + b)(a - b)$

Common error: $x^2 - 9 = (x - 3)^2$

Practice: (i) Factor: $25x^2 - 16y^2$ (ii) Rationalize the denominator: $\dfrac{2}{\sqrt{a} + b}$

Answer: (i) $(5x - 4y)(5x + 4y)$ (ii) $\dfrac{2(\sqrt{a} - b)}{a - b^2}$

A Great Website for More Detail: purplemath.com/modules/specfact.htm

8) Factoring Difference of Cubes: $a^3 - b^3 = (a - b)(a^2 + ab + b^2)$

Problem: (i) Factor: $8x^3 - 27$

(ii) Rationalize the denominator: $\dfrac{1}{\sqrt[3]{x} - 2}$

Solution:

(i) $8x^3 - 27 \overset{\boxed{\begin{array}{l} a = 2x \\ b = 3 \end{array}}}{=} (2x - 3)(4x^2 + 6x + 9)$

(ii) $\dfrac{1}{x^{1/3} - 2} \overset{\boxed{\begin{array}{l} \textbf{Use difference of cubes in reverse!} \\ a = x^{1/3} \;\; b = 2 \text{ so that } a^3 = x \text{ and } b^3 = 8 \end{array}}}{=} \dfrac{(x^{2/3} + 2x^{1/3} + 4)}{(x^{1/3} - 2)(x^{2/3} + 2x^{1/3} + 4)} = \dfrac{x^{2/3} + 2x^{1/3} + 4}{x - 8}$

Note: The order of the factors doesn't matter:

$a^3 - b^3 = (a - b)(a^2 + ab + b^2) = (a^2 + ab + b^2)(a - b)$

Common error: $x^3 - 9 = (x - 3)(x^2 + 3)$

Practice: (i) Factor: $x^6 - 64y^3$ (ii) Rationalize the denominator: $\dfrac{2}{a^{1/3} - b^{1/3}}$

Answer: (i) $x^6 - 64y^3 = (x^2 - 4y)(x^4 + 4x^2y + 16y^2)$

(ii) $\dfrac{2}{a^{1/3} - b^{1/3}} = \dfrac{2\,(a^{2/3} + a^{1/3}b^{1/3} + b^{2/3})}{a - b}$

A Great Website for More Detail: purplemath.com/modules/specfact2.htm

9) Factoring Sum of Cubes: $a^3 + b^3 = (a + b)(a^2 - ab + b^2)$

Problem: (i) Factor: $8x^3 + 27$

(ii) Rationalize the denominator: $\dfrac{1}{\sqrt[3]{x} + 2}$

Solution:

(i) $8x^3 + 27 \overset{\boxed{\substack{a = 2x \\ b = 3}}}{=} (2x + 3)(4x^2 - 6x + 9)$

(ii) $\dfrac{1}{x^{1/3} + 2} \overset{\boxed{\substack{\text{Use difference of cubes in reverse!} \\ a = x^{1/3} \ b = 2 \text{ so that } a^3 = x \text{ and } b^3 = 8}}}{=} \dfrac{(x^{2/3} - 2x^{1/3} + 4)}{(x^{1/3} + 2)(x^{2/3} - 2x^{1/3} + 4)} = \dfrac{x^{2/3} - 2x^{1/3} + 4}{x + 8}$

Note: The order of the factors doesn't matter:

$a^3 + b^3 = (a + b)(a^2 - ab + b^2) = (a^2 - ab + b^2)(a + b)$

Common error: $x^3 - 9 = (x - 3)(x^2 + 3)$

Practice: (i) Factor: $x^6 + 64y^3$ (ii) Rationalize the denominator: $\dfrac{2}{a^{1/3} + b^{1/3}}$

Answer: (i) $x^6 + 64y^3 = (x^2 + 4y)(x^4 - 4x^2y + 16y^2)$

(ii) $\dfrac{2}{a^{1/3} + b^{1/3}} = \dfrac{2\,(a^{2/3} - a^{1/3}b^{1/3} + b^{2/3})}{a + b}$

A Great Website for More Detail: purplemath.com/modules/specfact2.htm

10) Factoring $a^n - b^n = (a - b)(a^{n-1} + a^{n-2}b + a^{n-3}b^2 + a^{n-4}b^3 + \ldots + ab^{n-2} + b^{n-1})$

Problem: Factor: $x^5 - y^5$

Solution: $x^5 - y^5 = (x - y)(x^4 + x^3y + x^2y^2 + xy^3 + y^4)$

Note: This works for any positive integer n. Note the terms on the right are ALL positive. The exponents on a start at $n - 1$ and decrease to 0 and the exponents on b start at 0 and increase to $n - 1$.

Common error: $x^5 - y^5 = (x - y)(x^4 + y^4)$

Practice: Factor: $16x^4 - y^4$

Answer: $16x^4 - y^4 = (2x - y)(8x^3 + 4x^2y + 2xy^2 + y^3)$

A Great Website for More Detail: mathforum.org/library/drmath/view/55601.html

11) Factoring $a^n + b^n = (a + b)(a^{n-1} - a^{n-2}b + a^{n-3}b^2 - a^{n-4}b^3 + \ldots - ab^{n-2} + b^{n-1})$
and n **MUST BE ODD!**

Problem: Factor: $x^5 + y^5$

Solution: $x^5 + y^5 = (x + y)(x^4 - x^3y + x^2y^2 - xy^3 + y^4)$

Note: Expressions such as $x^2 + y^2$ and $x^4 + y^4$ (the exponent is even) DO NOT have a factor of $(x + y)$. This method only works when n is an ODD positive integer.

Common error: $x^5 + y^5 = (x + y)(x^4 + y^4)$

Practice: Factor: (i) $32x^5 + y^5$ where one factor is $(2x + y)$. (ii) $16x^4 + y^4$

Answer: (i) $32x^5 + y^5 = (2x + y)(16x^4 - 8x^3y + 4x^2y^2 - 2xy^3 + y^4)$

(ii) You cannot because the exponent is EVEN.

A Great Website for More Detail: en.wikipedia.org/wiki/Factorization

12) The Factor Theorem: Part 1

Problem: Factor the expression $x^3 - 4x^2 + x + 6$.

Solution: To find a factor, substitute $x = 1, -1, 2, -2, 3, -3, 6, -6$, in other words, divisors of 6, into the expression. The Factor Theorem tells us that if $x = a$ makes the expression equal to 0, then $x - a$ is a factor, that is, $x - a$ will divide evenly into $x^3 - 4x^2 + x + 6$. Then we can find the remaining factors without using the Factor Theorem again because the quotient will be a quadratic.

$x = 1$: $1 - 4 + 1 + 6 = 4 \neq 0$
$x = -1$: $-1 - 4 - 1 + 6 = 0 = 0$

Therefore $x + 1$ is a factor. Now we divide $x + 1$ into $x^3 - 4x^2 + x + 6$ as the first step to finding the other factors.

$$
\begin{array}{r}
x^2 - 5x + 6 \\
x + 1 \overline{) x^3 - 4x^2 + x + 6} \\
-(x^3 + x^2) \\
\hline
-5x^2 + x + 6 \\
-(-5x^2 - 5x) \\
\hline
6x + 6 \\
-(6x + 6) \\
\hline
0
\end{array}
$$

$\boxed{\text{The quadratic is easy to factor!}}$

$\therefore x^3 - 4x^2 + x + 6 = (x + 1)(x^2 - 5x + 6) \qquad = \qquad (x + 1)(x - 2)(x - 3)$

Note: Polynomial division is just like ordinary division. For example,

$$
\begin{array}{r}
9 \\
3 \overline{) 27} \\
-(27) \\
\hline
0
\end{array}
$$

Common error: Not properly subtracting to get the remainder at each step in the polynomial division.

Practice: Use the Factor Theorem to factor $x^3 - 7x^2 + 16x - 12$

Answer: $x^3 - 7x^2 + 16x - 12 = (x - 2)^2 (x - 3)$

A Great Website for More Detail: purplemath.com/modules/factrthm.htm

13) The Factor Theorem: Part 2

Problem: Find all the rational roots of $2x^3 - 5x^2 - 4x + 3$.

Solution: By the Factor Theorem, all rational roots must be of the form $\dfrac{a}{b}$,

where a is a divisor of 3 and b is a divisor of 2. So we will substitute

$x = 1, -1, 3, -3, \dfrac{1}{2}, -\dfrac{1}{2}, \dfrac{3}{2}, -\dfrac{3}{2}$ in the original expression. If $x = r$

makes the expression equal to 0, then $x - r$ is a factor and r is a root. Since we have cubic, there are at most three rational factors.

$x = 1$: $2 - 5 - 4 + 3 = -4 \neq 0$
$x = -1$: $-2 - 5 + 4 + 3 = 0$
$x = 3$: $54 - 45 - 12 + 3 = 0$
$x = -3$: $-54 - 45 + 12 + 3 \neq 0$

$x = \dfrac{1}{2}$: $\dfrac{1}{4} - \dfrac{5}{4} - 2 + 3 = 0$

Since we have three roots and the original expression is a cubic polynomial, we are finished.

The rational roots (and in fact all the roots) are -1, 3, and $\dfrac{1}{2}$

Note: In factored form, $2x^3 - 5x^2 - 4x + 3 = 2(x + 1)(x - 3)\left(x - \dfrac{1}{2}\right) = (x + 1)(x - 3)(2x - 1)$

Common error: Here, it is very easy to make arithmetic errors when substituting

numbers like $\dfrac{1}{2}$ into the expression.

Practice: Use the Factor Theorem to find the rational roots of $2x^3 - 3x^2 - x - 2$.

Answer: $x = 2$; the other two roots are complex numbers.

$2x^3 - 3x^2 - x - 2 = (2x^2 + x + 1)(x - 2)$

A Great Website for More Detail: purplemath.com/modules/factrthm.htm

14) Polynomial Division

Problem: Divide $x^3 - 5x^2 + x + 6$ by $x - 4$; express your answer $\dfrac{x^3 - 5x^2 + x + 6}{x - 4}$ both in the form $\dfrac{x^3 - 5x^2 + x + 6}{x - 4} = \text{Quotient} + \dfrac{\text{Remainder}}{\text{Divisor}}$

and $x^3 - 5x^2 + x + 6 = \text{Quotient} \times \text{Divisor} + \text{Remainder}$.

Solution:

$$
\begin{array}{r}
x^2 - x - 3 \\
x - 4 \,\overline{)\, x^3 - 5x^2 + x + 6} \\
-(x^3 - 4x^2) \\
\hline
-x^2 + x + 6 \\
-(-x^2 + 4x) \\
\hline
-3x + 6 \\
-(-3x + 12) \\
\hline
-6
\end{array}
$$

Therefore, $\dfrac{x^3 - 5x^2 + x + 6}{x - 4} = x^2 - x - 3 - \dfrac{6}{x - 4}$

and $x^3 - 5x^2 + x + 6 = (x - 4)(x^2 - x - 3) - 6$

Note: Compare $\dfrac{28}{3} = \text{Quotient} + \dfrac{\text{Remainder}}{\text{Divisor}} = 9 + \dfrac{1}{3}$ with

$\dfrac{x^3 - 5x^2 + x + 6}{x - 4} = \text{Quotient} + \dfrac{\text{Remainder}}{\text{Divisor}} = x^2 - x - 3 - \dfrac{6}{x - 4}$

Common error: Not properly subtracting to get the remainder at each step in the polynomial division.

Practice: Simplify $\dfrac{x^3 - 2x^2 - 3x + 3}{x + 2}$ and express your answer as in the example.

Answer: $\dfrac{x^3 - 2x^2 - 3x + 6}{x + 2} = x^2 - 4x + 5 - \dfrac{7}{x + 2}$

$x^3 - 2x^2 - 3x + 3 = (x^2 - 4x + 5)(x + 2) - 7$

A Great Website for More Detail: purplemath.com/modules/factrthm.htm

15) Solving for the Roots of a Polynomial

Problem: Solve: (i) $x^2 - 5x + 6 = 0$ (ii) $x^2 - 5x + 1 = 0$ (iii) $x^3 - 7x = -6$

Solution: (i) Factoring, $x^2 - 5x + 6 = (x - 3)(x - 2) = 0$ and so $x = 2$ or $x = 3$.

(ii) Using the quadratic formula with $a = 1$ and $b = -5$ and $c = 1$, we get

$$x \overset{\boxed{\frac{-b \pm \sqrt{b^2 - 4ac}}{2a}}}{=} \frac{-(-5) \pm \sqrt{(-5)^2 - 4(1)(1)}}{2(1)} = \frac{5 \pm \sqrt{21}}{2} \text{, that is, the roots are}$$

$$x = \frac{5 + \sqrt{21}}{2} \text{ and } x = \frac{5 - \sqrt{21}}{2}.$$

(iii) Apply the Factor Theorem to $P(x) = x^3 - 7x + 6$:

$$P(1) = 0 \quad P(-1) = 12 \quad P(2) = 0 \quad P(-2) = 12 \quad P(3) = 12 \quad P(-3) = 0$$

We can stop even though we haven't checked $P(6)$ nor $P(-6)$ (remember that we check all divisors of 6) because we KNOW a cubic polynomial has exactly three roots. So the roots are 1, 2, and -3.

Note 1: $P(x) = a_n x^n + a_{n-1} x^{n-1} + \dots + a_1 x + a_0$ has roots $r_1, r_2, r_3, \dots,$ and $r_n \Leftrightarrow$
$P(x) = a_n(x - r_1)(x - r_2)(x - r_3)\dots(x - r_n)$

Note 2: $P(x) = a_n x^n + a_{n-1} x^{n-1} + \dots + a_1 x + a_0$ has roots $r_1, r_2, r_3, \dots,$ and $r_n \Leftrightarrow$
the x intercepts of $y = P(x)$ are the REAL numbers in the set $\{r_1, r_2, r_3, \dots, r_n\}$

Common errors: (i) Getting the signs wrong when factoring a quadratic. (ii) Substituting incorrectly when using the quadratic formula. (iii) Not evaluating $P(a)$ correctly when using the Factor Theorem.

Practice: Solve: (i) $x^2 + 11x + 30 = 0$ (ii) $x^2 - 2x + 2 = 0$ (iii) $x^3 - x^2 + x - 1 = 0$

Answer: (i) $-5, -6$ (ii) $1 + i, 1 - i$ (iii) $1, i, -i$ (Remember, $i = \sqrt{-1}$.)

A Great Website for More Detail: oakroadsystems.com/math/polysol.htm

16) Solving "Factored" Inequalities: Numerators Only

Problem: Solve: (i) $(x + 1)(x - 1)(x - 2) \geq 0$ (ii) $(x + 1)^2 x(x - 1)^3 > 0$

Solution: (i) The only x values where the expression can change from $+$ to $-$ or $-$ to $+$ are -1, 1, and 2. So we analyze the sign of each factor in the expression on a number line using these values.

$$
\begin{array}{cccc}
x < -1 & -1 < x < 1 & 1 < x < 2 & x > 2 \\
(-)(-)(-) & (+)(-)(-) & (+)(+)(-) & (+)(+)(+)
\end{array}
$$

$$
\underset{-\qquad -1 \qquad + \qquad 1 \qquad - \qquad 2 \qquad +}{\xrightarrow{\hspace{6cm}}} x
$$

The solution in interval notation is $[-1,1] \cup [2,\infty)$.

(ii) The only x values where the expression can change from $+$ to $-$ or $-$ to $+$ are -1, 0, and 1. **However,** the exponent on $x + 1$ is even: $(x + 1)^2 \geq 0$ always!

$$
\begin{array}{cccc}
x < -1 & -1 < x < 0 & 0 < x < 1 & x > 1 \\
(+)(-)(-) & (+)(-)(-) & (+)(+)(-) & (+)(+)(+)
\end{array}
$$

$$
\underset{+ \qquad -1 \qquad + \qquad 0 \qquad - \qquad 1 \qquad +}{\xrightarrow{\hspace{6cm}}} x
$$

The solution in interval notation is $(-\infty,-1) \cup (-1,0) \cup (1,\infty)$.

Note 1: If one of the factors is, for example, $(x + 1)^{1/3}$, then the term will change sign at $x = -1$. If the factor were $(x + 1)^{2/3}$, it would not change sign.

Note 2: What is the geometric significance of this kind of problem? In (i), you have found where the graph of $y = (x + 1)(x - 1)(x - 2)$ is on or above the x axis.

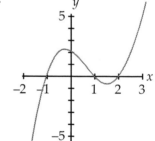

Common error: Getting the signs wrong when the exponent on a factor is even.

Practice: (i) $-(x - 2)^3 x(x - 3)^{1/5} \geq 0$ (ii) $-(x - 2)^4 x(x - 3)^{2/5} \geq 0$

Answer: (i) $(-\infty,0] \cup [2,3]$ (ii) $(-\infty,0] \cup \{2,3\}$

Note: Because the expression in (ii) equals 0 at $x = 2$ and $x = 3$ (as well as $x = 0$), 2 and 3 are part of the solution.

A Great Website for More Detail: purplemath.com/modules/ineqsolv.htm

17) Solving "Factored" Rational Inequalities:

Problem: Solve: (i) $\dfrac{(x+1)(x-2)}{x-1} \geq 0$ (ii) $\dfrac{(x+1)^2(x-1)^3}{x} > 0$

Solution: (i) The only x values where the expression can change from + to − or − to + are −1, 1, and 2. So we analyze the sign of each factor in the expression on a number line using these values.

$$
\begin{array}{cccc}
x < -1 & -1 < x < 1 & 1 < x < 2 & x > 2 \\
(-)(-)(-) & (+)(-)(-) & (+)(+)(-) & (+)(+)(+)
\end{array}
$$

$$
\underset{\;\;-\quad\;\; -1 \quad\; + \quad\; 1 \quad\; - \quad\; 2 \quad\; +}{\rule{8cm}{0.4pt}} \; x
$$

The question asks for values of x where the expression is "≥ 0". So values of x where the numerator is 0 work. Values of x where the denominator is 0 **do not!** We inclue $x = -1$ and $x = 2$ but not $x = 1$. The solution in interval notation is $[-1,1) \cup [2, \infty)$.

(ii) The only x values where the expression can change from + to − or − to + are −1, 0, and 1. **However,** the exponent on x + 1 is even: $(x+1)^2 \geq 0$ always!

$$
\begin{array}{cccc}
x < -1 & -1 < x < 0 & 0 < x < 1 & x > 1 \\
(+)(-)(-) & (+)(-)(-) & (+)(+)(-) & (+)(+)(+)
\end{array}
$$

$$
\underset{\;\;+\quad\;\; -1 \quad\; + \quad\; 0 \quad\; - \quad\; 1 \quad\; +}{\rule{8cm}{0.4pt}} \; x
$$

The question asks for values of x where the expression is "> 0". So values of x where the numerator is 0 **do not work.** Values of x where the denominator is 0 **never work!** We exclude $x = -1, 0, 1$. The solution in interval notation is $(-\infty,-1) \cup (-1,0) \cup (1, \infty)$.

Note 1: If one of the factors is, for example, $(x+1)^{1/4}$, then $x \geq -1$ because you can't take an even root of a negative number.

Note 2: If one of the factors is, for example, $(x+1)^{1/3}$, then the term will change sign at $x = -1$. If the factor were $(x+1)^{2/3}$, it would not change sign.

Common error: Getting the signs wrong when the exponent on a factor is even.

Practice: (i) $\dfrac{-(x-3)^{1/5}}{x(x-2)^3} \geq 0$ (ii) $\dfrac{-(x-3)^{2/5}}{(x-2)^4 x} \geq 0$

Answer: (i) $(-\infty,0) \cup (2,3]$ (ii) $(-\infty,0) \cup \{3\}$

A Great Website for More Detail: purplemath.com/modules/ineqrtnl.htm

18) Completing the Square

Problem: Complete the square: (i) $x^2 - 8x + 25$ (ii) $3x^2 + 36x - 17$ (iii) $-2x^2 + 3x + 1$

Solution: Remember $(x + a)^2 = x^2 + 2ax + a^2$

(i) $x^2 - 8x + 25$

$\begin{array}{c} 2a = -8 \\ \therefore a = -4 \\ \text{and } a^2 = 16 \end{array}$ $\boxed{2a}$ $\boxed{a^2}$ We added 16 so we subtract 16 to compensate! $\boxed{a}$

$= x^2 - 8x + 16 + 25 \qquad -16 \qquad = (x - 4)^2 + 9$

(ii) $3x^2 + 36x - 17 = 3(x^2 + 12x) - 17$

$\begin{array}{c} 2a = 12 \\ \therefore a = 6 \\ \text{and } a^2 = 36 \end{array}$

$= 3(x^2 + 12x + 36) - 17$

We added $3 \times 36 = 108$ so we subtract 108 to compensate!

$-108 \qquad = 3(x + 6)^2 - 125$

(iii) $-2x^2 + 3x + 1 = -2(x^2 - \dfrac{3}{2}x) + 1$

$\begin{array}{c} 2a = -\dfrac{3}{2} \quad \therefore a = -\dfrac{3}{4} \\ \text{and } a^2 = \dfrac{9}{16} \end{array}$

$= -2(x^2 - \dfrac{3}{2}x + \dfrac{9}{16}) + 1$

Because of the -2 outside the bracket, we subtracted $2\dfrac{9}{16} = \dfrac{9}{8}$ so we add $\dfrac{9}{8}$ to compensate

$+ \dfrac{9}{8} \qquad = -2\left(x - \dfrac{3}{4}\right)^2 + \dfrac{17}{8}$

Note: When completing the square on the expression $Ax^2 + Bx + C$, start this way:

$Ax^2 + Bx + C = A \left(x^2 + \dfrac{B}{A}x \right) + C$, that is, factor the A from the x^2 and x terms, but not the constant.

Common error: Not "compensating" correctly

Practice: (i) $x^2 - 4x - 3$ (ii) $-3x^2 + 5x + 1$

Answer: (i) $(x - 2)^2 - 7$ (ii) $-3\left(x - \dfrac{5}{6}\right)^2 + \dfrac{37}{12}$

A Great Website for More Detail: purplemath.com/modules/sqrquad.htm

19) Adding and Subtracting Rational Expressions

Problem: By getting a common denominator, simplify the following expressions:

(i) $\dfrac{1}{3x} - \dfrac{1}{2y} + \dfrac{1}{6z}$ (ii) $\dfrac{2x+1}{x-1} - \dfrac{x+1}{x+2} - \dfrac{5x+4}{x^2+x-2}$

Solution: (i) $\dfrac{1}{3x} - \dfrac{1}{2y} + \dfrac{1}{6z}$

$\boxed{\text{Get the lowest common denominator!}}$
$= \dfrac{2yz}{6xyz} - \dfrac{3xz}{6xyz} + \dfrac{xy}{6xyz}$ $\boxed{\text{Simplify the numerator.}}$ $= \dfrac{2yz - 3xz + xy}{6xyz}$

(ii) $\dfrac{2x+1}{x-1} + \dfrac{x+1}{x+2} - \dfrac{5x+4}{x^2+x-2}$

$\boxed{\text{Factor the denominators.}}$
$= \dfrac{2x+1}{x-1} - \dfrac{x+1}{x+2} - \dfrac{5x+4}{(x-1)(x+2)}$

$\boxed{\text{Get the lowest common denominator.}}$
$= \dfrac{(2x+1)(x+2)}{(x-1)(x+2)} - \dfrac{(x+1)(x-1)}{(x-1)(x+2)} - \dfrac{5x+4}{(x-1)(x+2)}$

$\boxed{\text{Expand the numerator.}}$
$= \dfrac{2x^2 + 5x + 2 - (x^2 - 1) - (5x+4)}{(x-1)(x+2)}$

$\boxed{\text{Now simplify the numerator.}}$
$= \dfrac{x^2 - 1}{(x-1)(x+2)}$

$\boxed{\begin{array}{c}\text{Check for and divide out} \\ \text{any further common factors.}\end{array}}$
$= \dfrac{\cancel{(x-1)}(x+1)}{\cancel{(x-1)}(x+2)} = \dfrac{x+1}{x+2}$

Note: Adding and subtracting expressions use **exactly** the same methods as adding and subtracting ordinary numerical fractions!

Common error: Not using the lowest common denominator which leads to a more complicated expression and more chances of making a mistake!

Practice: (i) $\dfrac{1}{2a} - \dfrac{1}{3b} + \dfrac{1}{6ab}$ (ii) $\dfrac{1}{x^2} + \dfrac{x-2}{x(x+2)} - \dfrac{x}{(x+2)^2}$

Answer: (i) $\dfrac{9b - 2a + 5}{6ab}$ (ii) $\dfrac{x^2 + 4}{x^2(x+2)^2}$

A Great Website for More Detail: purplemath.com/modules/rtnladd.htm

20) Multiplying and Dividing Rational Expressions

Problem: Simplify the following rational expressions:

(i) $\dfrac{(x^2-16)}{(x-4)^3} \times \dfrac{(x^2-4x)^2}{x^3+64}$ (ii) $\dfrac{x^2+5xy+4y^2}{x^2+4xy+4y^2} \div \dfrac{x^2+xy}{x^2+2xy}$

Solution: (i) $\dfrac{(x^2-16)}{(x-4)^3} \times \dfrac{(x^2-4x)^2}{x^3+64}$

$= \dfrac{\cancel{(x-4)}(x+4)}{\cancel{(x-4)^3}} \times \dfrac{x^2\cancel{(x-4)^2}}{\cancel{(x+4)}(x^2-4x+16)} = \dfrac{x^2}{x^2-4x+16}$

(ii) $\dfrac{x^2+5xy+4y^2}{x^2+4xy+4y^2} \div \dfrac{x^2+xy}{x^2+2xy}$

$\boxed{\text{Factor!}}$
$= \dfrac{(x+4y)(x+y)}{(x+2y)^2} \div \dfrac{x(x+y)}{x(x+2y)}$

$\boxed{\begin{array}{c}\text{Invert and}\\ \text{multiply}\end{array}}$
$= \dfrac{(x+4y)(x+y)}{(x+2y)^2} \times \dfrac{x(x+2y)}{x(x+y)}$

$\boxed{\begin{array}{c}\text{Divide out}\\ \text{common factors}\end{array}}$
$= \dfrac{x+4y}{x+2y}$

Note: When you multiply or divide rational expressions, you use exactly the same methods as adding and subtracting ordinary numerical fractions!

Common errors: Students often divide out "terms" instead of "factors". For example, in the original question (i), x^2 is a term, while (x^2-16) is a factor.

Practice: Simplify the following rational expressions:

(i) $\dfrac{(x^2+7x+6)^2}{(x^2+12x+36)(x^2-1)} \times \dfrac{x^3-1}{x^2+x+1}$ (ii) $\dfrac{(z^3+5z^2)^3}{z^2+10z+25} \div \dfrac{z^8(z^3+125)}{z^2-5z+25}$

Answer: (i) $x+1$ (ii) $\dfrac{1}{z^2}$

A Great Website for More Detail: purplemath.com/modules/rtnlmult.htm

ANSWERS
Additional practice and web references

Part IV: Geometry

Part IV: Geometry

1) Pythagorean Theorem

Problem: In the following diagrams find the value of the unknowns:

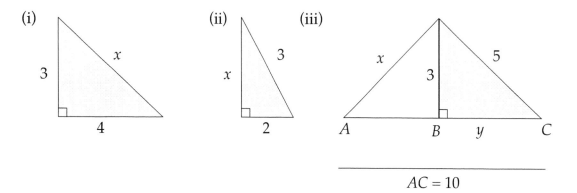

Solution: (i) $x^2 = 3^2 + 4^2 = 25$ ∴ $x = 5$

(ii) $3^2 = 2^2 + x^2$ ∴ $x^2 = 9 - 4 = 5$ ∴ $x = \sqrt{5}$

(iii) Let $BC = y$. Then $5^2 = 25 = y^2 + 3^2$. ∴ $y^2 = 25 - 9 = 16$ ∴ $y = 4$

$AB = 10 - y = 10 - 4 = 6$ Now we can find x: $x^2 = 3^2 + 6^2 = 45$ ∴ $x = \sqrt{45} = 3\sqrt{5}$

Note: The important point is that if the length of any two sides of a right angled triangle is known we can determine the third side using the Pythagorean Theorem.

Common error: $x^2 = a^2 + b^2$ ∴ $x = a + b$

Practice:
Find x in the
following:

(i)   (ii)

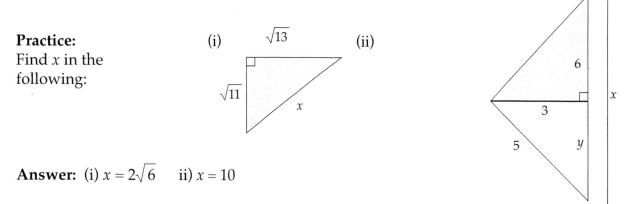

Answer: (i) $x = 2\sqrt{6}$ ii) $x = 10$

A Great Website for More Details: purplemath.com/modules/perimetr3.htm

2) Angles in a Triangle

Problem: Find the values of angles x and y (in degree measure) from the following diagrams:

(i)   (ii)

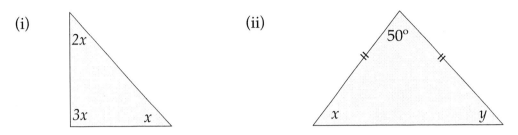

Solution: (i) $x + 2x + 3x = 180°$ ∴ $6x = 180°$ ∴ $x = 30°$

(ii) $x + y + 50° = 180°$ But $x = y$ (Isosceles Triangle!)

So $2x + 50° = 180°$ and $x = y = 65°$.

Note: Once you know any two angles of a triangle you know all three.

Common error: Making a mistake when solving the equation for the unknown.

Practice: Find x in each of the following diagrams:

(i)   (ii)

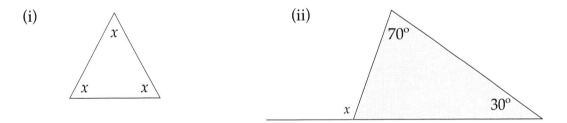

Answer: (i) $x = 60°$ (ii) $x = 100°$

A Great Website for More Details: mathopenref.com / triangleinternalangles.html

3) The Parallel Line Theorem

Problem: Find the values in degrees of x and y in the following diagrams:

(i)   (ii)

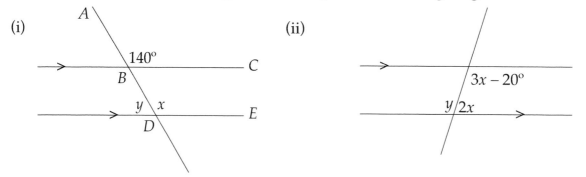

Solution: i) $\angle ABC = \angle BDE = x$ (corresponding angles)

$\therefore x = 140°$ Also $x + y = 180°$ and so $y = 40°$

(ii) $y = 3x - 20°$ (alternate angles) Also, $2x + y = 180°$

$\therefore 2x + 3x - 20° = 180° \Rightarrow 5x = 200° \Rightarrow x = 40°$

So $x = 40°$ and $y = 3(40°) - 20° = 100°$

Note: When two lines are parallel, the sum of the interior angles on the same side of the transversal sum to 180°. So we could have solved (ii) by writing $2x + 3x - 20° = 180°$ and then finding first x and then y.

Common error: Misidentification of corresponding and/or alternate angles.

Practice: Find the values in degrees of x and y:

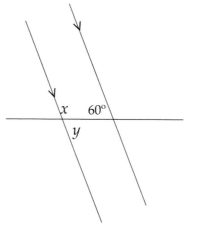

Answer: $x = 120°$, $y = 60°$

A Great Website for More Detail:

sonoma.edu/users/w/wilsonst/Papers/Geometry/parallel/default.html

4) Congruent Triangles

Problem: Triangles I and II are congruent. Name the congruent triangles so that the vertices "correspond" and determine the values x, y, u, and v.

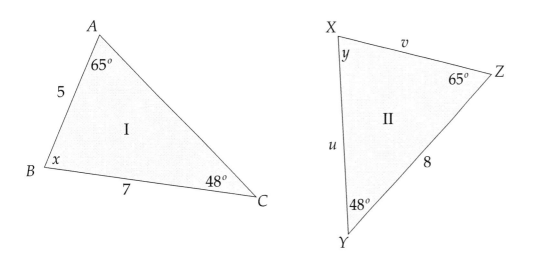

Solution: Note that the two triangles are congruent by AAS: $\triangle ABC \cong \triangle ZXY$

$x = 180° - 45° - 65° = 67° = y$ $\qquad$ $\boxed{u = XY = BC}$ $\quad u = 7$ $\qquad$ $\boxed{v = XZ = BA}$ $\quad v = 5$

Note: When two congruent triangles are superimposed, every corresponding component matches up: sides, angles, area, perimeter, everything. Congruency can be determined by SSS, SAS, ASA, and AAS but not by SSA nor by AAA!

Common error: $\triangle ABC \cong \triangle ZYX$ This is wrong because the vertices do not correspond.

Practice: The two triangles below are congruent. Name the congruent triangles so that the vertices "correspond" and determine the values a, b, c, and d.

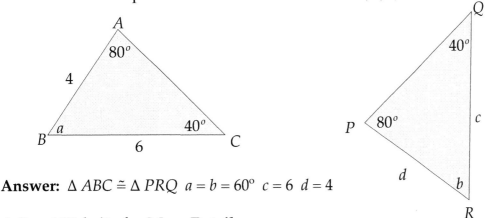

Answer: $\triangle ABC \cong \triangle PRQ$ $\quad a = b = 60°$ $\quad c = 6$ $\quad d = 4$

A Great Website for More Detail:
sonoma.edu/users/w/wilsonst/Papers/Geometry/parallel/default.html

5) Similar Triangles

Problem: Triangles I and II are similar. Name the similar triangles so that the vertices "correspond" and determine the values x, y, and z.

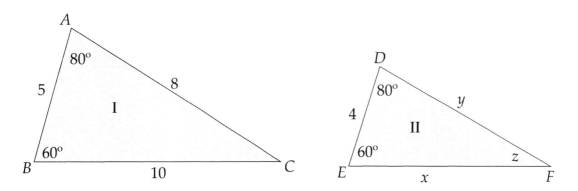

Solution: Note that the two triangles are similar by AA: $\triangle ABC \sim \triangle DEF$

$z = 180° - 80° - 60° = 40°$ By similar triangles, $\dfrac{x}{10} = \dfrac{y}{8} = \dfrac{4}{5}$

Solving gives $x = 8$ and $y = \dfrac{32}{5}$.

Note: Similarity can be determined by SSS, SAS, and AA (which is the same as AAA!) Here, when we use S, we are referring not to equality of sides but to the ratio of corresponding sides.

Common error: $\triangle ABC \sim \triangle EDF$ This is wrong because the vertices do not correspond.

Practice: The two triangles below are similar. Name the similar triangles so that the vertices "correspond" and determine the values x and y.

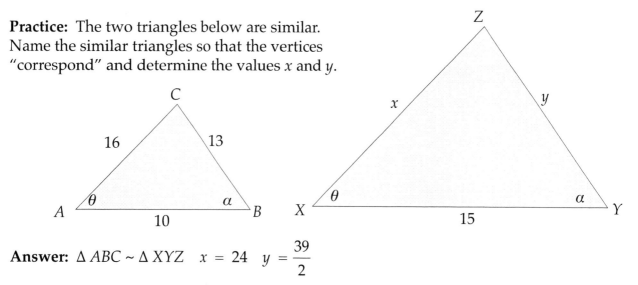

Answer: $\triangle ABC \sim \triangle XYZ$ $x = 24$ $y = \dfrac{39}{2}$

A Great Website for More Detail:
sonoma.edu/users/w/wilsonst/Papers/Geometry/parallel/default.html

6) Area of a Triangle

Problem: Find the area of Δ *ABC* in each of the following:

(i)

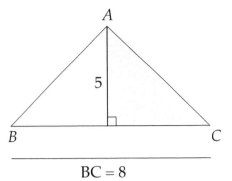

BC = 8

(ii)

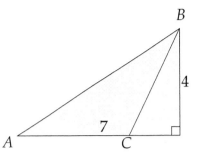

Solution: (i) Area = $\frac{1}{2}$ base × height = $\frac{1}{2}$ × 8 × 5 = 20 square units.

(ii) Area = $\frac{1}{2}$ base × height = $\frac{1}{2}$ × 7 × 4 = 14 square units

Note: Choose one side to be the base. Then the height is the **length of the perpendicular** from the vertex opposite to the base. Sometimes the foot of the **perpendicular** is on an extension of the base as in diagram (ii).

Common error: Not realizing the perpendicular height from *B* to *AC* is on an extension of *AC*.

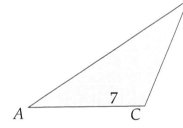

Practice: Find the area of Δ *ABC*.
(Hint: first use the Pythagorean Theorem to find the perpendicular height.)

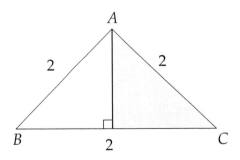

Answer: Area = $\sqrt{3}$ sq. units.

A Great Website for More Detail:
en.wikipedia.org/wiki/Triangle#Computing_the_area_of_a_triangle

7) Area and Circumference of a Circle

Problem: (i) Find the circumference and area of a circle of radius 8 cm.
(ii) Find the circumference and area of a circle of diameter 8 m.
(iii) If the area of a circle is 16π, find the radius.

Solution: (i) Radius $r = 8$ cm
Circumference C $= 2\pi r = 2\pi \times 8 = 16\pi$ cm and area A $= \pi r^2 = \pi 8^2 = 64\pi$ cm^2

(ii) Diameter $d = 8$ C $= \pi d = \pi \times 8 = 8\pi$ m and area A $= \pi \times \dfrac{d^2}{4} = \pi \times \dfrac{8^2}{4} = 16\pi$ m^2.

(iii) A $= \pi r^2 = 16\pi$ Then $r^2 = 16$ and $r = 4$ m.

Note: Take any circle – ANY CIRCLE! – and wrap copies of the radius around the semi-circle. How many copies will it take? $\pi = 3.14159...$

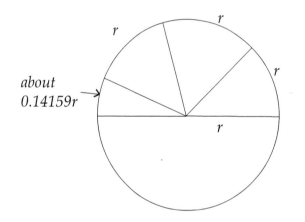

Common error: Using the diameter value d in the formulas $2\pi r$ and πr^2 ; similarly using the radius in the formulas πd and $\dfrac{\pi d^2}{4}$

Practice: Find the circumference and area of a circle of diameter 12 units.

Answer: C $= 12\pi$ units and A $= 36\pi$ units^2

A Great Website for More Detail: mathforum.org/dr.math/faq/faq.pi.html

8) Arc Length and Area of a Sector of a Circle

Problem: (i) State the arc length s and the area A of the sector of this circle. Assume θ is in **radian** measure.

(ii) In the circle below, find the arc length s and area A.

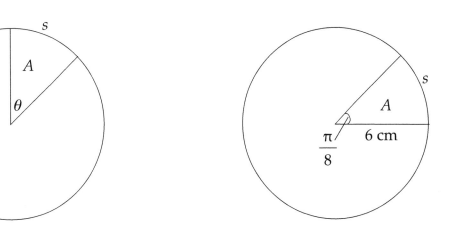

Solution: (i) Arc length $s = r\theta$ units and $A = \dfrac{1}{2} r^2\theta$ units^2. (Remember: the angle must be measured in radians.)

(ii) $s = r\theta = 6 \times \dfrac{\pi}{8} = \dfrac{3\pi}{4}$ cm; $A = \dfrac{1}{2} r^2\theta = \dfrac{1}{2}(6)^2 \left(\dfrac{\pi}{8} \right) = \dfrac{9\pi}{4}$ cm^2

Note: In the correct use of the formulas in (i), θ must be in radians. If θ is given in degrees convert to radians using this conversion formula:

angle in radians $= \theta \times \dfrac{\pi}{180°}$

(VERY) Common error: Using θ in degrees in the formulas in (i).

Practice: In the circle to the right, find the length s corresponding sector area A.

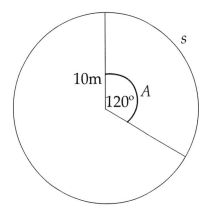

(Hint: $120° = \dfrac{2\pi}{3}$ radians)

Answer: $s = \dfrac{20\pi}{3}$ m; $A = \dfrac{100\pi}{3}$ m^2

A Great Website for More Details:
worsleyschool.net/science/files/sector/calculations.html

9) Volume of a Sphere, Box, Cone, Cylinder

Problem: Find the volume V of
(i) a sphere of radius $r = 3$ cm
(ii) a rectangular box with length $l = 8$ cm, width $w = 5$ cm, and height $h = 50$ cm
(iii) a right circular cone with height $h = 3$ cm and base radius $r = 0.04$ m
(iv) a circular cylinder with height $h = 0.2$ m and radius $r = 4$ cm.

Solution: i) $V = \dfrac{4}{3}\pi r^3 = \dfrac{4}{3}\pi(3^3) = 36\,\pi$ cm³

(ii) $V = l \times w \times h = 8 \times 5 \times 50 = 2000$ cm³

(iii) Get common units! $r = 0.04$ m $= 4$ cm $\therefore V = \dfrac{1}{3}\pi r^2 h = \dfrac{1}{3}\pi \times 4^2 \times 3 = 16\pi$ cm³

(iv) $h = 0.2$ m $= 20$ cm $\therefore V = \pi r^2 h = 4^2 \times 20 = 320\pi$ cm³

Note: You have used a lot of volume formulas (like the cone and the sphere) as early as Grade 3. In calculus, you finally get to **PROVE** them.

Common error: Not using a common measurement unit (such as cm) for all the variables in the volume formula.

Practice: Find the volume V of

(i) a sphere of radius 2 cm

(ii) a box of length 0.3 m, width 7 cm, and height 3 cm

(iii) a cone of height 8 cm and radius 4 cm

(iv) a circular cylinder of height 12 cm and base radius 3 cm.

Answers: i) $\dfrac{32\pi}{3}$ cm³   ii) 630 cm³   iii) $\dfrac{128\pi}{3}$ cm³   iv) 108π cm³

A Great Website for More Detail:
en.wikipedia.org/wiki/Volume

10) Angles of a Polygon

Problem: i) Find the sum of the interior angles of a pentagon.

(ii) If all the interior angles of a pentagon are equal how much is each interior angle?

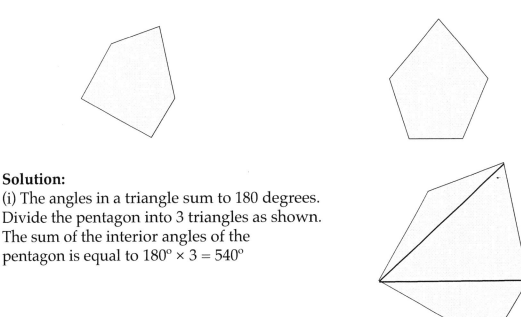

Solution:

(i) The angles in a triangle sum to 180 degrees. Divide the pentagon into 3 triangles as shown. The sum of the interior angles of the pentagon is equal to $180° \times 3 = 540°$

(ii) Since the 5 interior angles are equal each interior angle is equal to $\dfrac{540°}{5} = 108°$.

Note: By drawing lines from one vertex to each of the others in a polygon, you subdivide a polygon with 4 sides (a quadrilateral) into 2 triangles, a polygon with 5 sides (a pentagon) into 3 triangles, and, in general, a polygon with n sides (an "n-gon") into $n - 2$ triangles. The sum of the interior angles of an n-sided polygon is equal to $180(n - 2)$ degrees.

Common error: Dividing $180(n - 2)$ degrees by n to determine the size of an interior angle of an n-sided polygon. This is only correct when we know that all the n interior angles of the n-sided polygon are equal, that is, the polygon is "regular".

Practice: (i) Find the sum of the interior angles of a polygon with 9 sides.
(ii) Find the interior angle of a regular 9-gon.

Answers: (i) 1260° (ii) 140°

A Great Website for More Detail: regentsprep.org/regents/math/poly/LPoly1.htm

ANSWERS
Additional practice
and web references

Part V: Basic Graphs

Part V: Basic Graphs

1) Graphing $y = x^n$, $n \in \mathbf{N}$

Problem: (i) Graph the curves $y = x^2$ and $y = x^4$ on the same set of axes.
(ii) Graph the curves $y = x^3$ and $y = x^5$ on the same set of axes.

Solution: (i)　　　　　　　　　　　　　　　(ii)

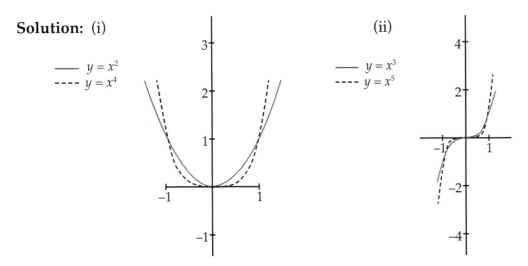

Note: When n is even, $y = x^n$ is symmetric in the y axis. That is why graphs symmetric in the y axis are called **even functions**. When n is odd, $y = x^n$ is symmetric in the origin. That is why graphs symmetric in the origin are called **odd functions**.

Common error: Most of us realize $y = x^4$ is above $y = x^2$ when $x > 1$ but do not realize it is BELOW $y = x^2$ when $0 < x < 1$.

Practice: Graph $y = x^2$ and $y = x^3$ on the same set of axes.

Answer:

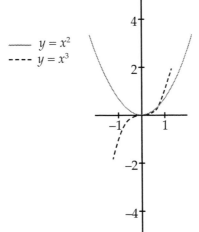

A Great Website for More Detail:
ugrad.math.ubc.ca/coursedoc/math100/notes/zoo/powers.html (requires JAVA)

2) Graphing $y = x^{m/n}$, $m, n \in \mathbb{N}$, n Odd, and m/n is a Reduced Fraction

Problem: (i) Graph the curves $y = x^{1/3}$ and $y = x^{2/3}$ on the same set of axes.
(ii) Graph the curves $y = x^{4/3}$ and $y = x^{5/3}$ on the same set of axes.

Solution:
(i)

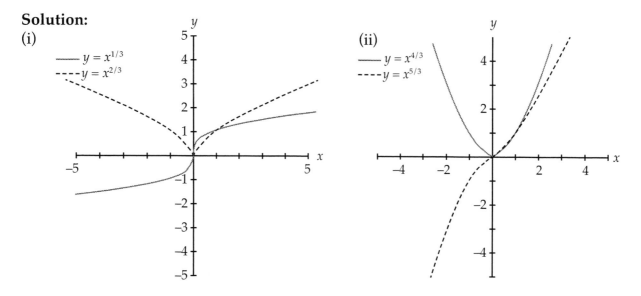

Note: When $0 < m/n < 1$, the graph of $y = x^{m/n}$ looks like the square root or cube root function. When $m/n > 1$, the graph of $y = x^{m/n}$ looks like the parabola or the cubic.

Common error: Mixing up when $y = x^{m/n}$ is always positive and when it is sometimes $+$ and sometimes $-$.

Practice: Graph $y = x^{3/5}$ and $y = x^{5/3}$ on the same set of axes.

Answer:

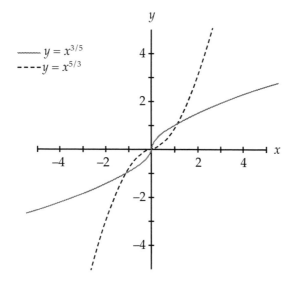

A Great Website for More Detail: wmueller.com/precalculus/families/pwrfrac.html

3) Graphing $y = x^{m/n}$, m, $n \in$ **N**, n Even, and m/n is a Reduced Fraction

Problem: Graph the curves $y = x^{1/2}$ and $y = x^{3/2}$ on the same set of axes.

Solution:

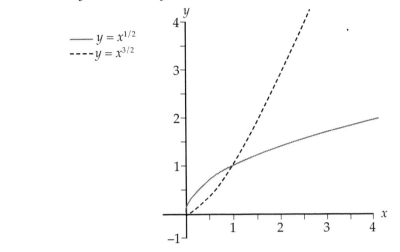

Note: The domain of both of these functions is $\{x \in$ **R** $\mid x \geq 0\}$. When the exponent is between 0 and 1, the graph is concave down. When it is greater than 1, it is concave up.

Common error: Forgetting that x must be ≥ 0.

Practice: Graph $y = x^{1/4}$ and $y = x^{5/4}$ on the same set of axes.

Answer:

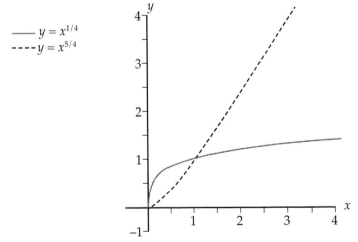

A Great Website for More Detail:

ugrad.math.ubc.ca/coursedoc/math100/notes/zoo/powers2.html (requires JAVA)

4) Graphing $y = x^{-n} = \dfrac{1}{x^n}$, $n \in \mathbf{N}$

Problem: Graph the curves $y = x^{-1} = \dfrac{1}{x}$ and $y = x^{-2} = \dfrac{1}{x^2}$ on the same set of axes.

Solution:

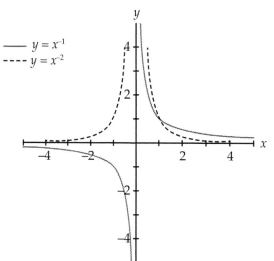

Notes: First, we know $x \neq 0$. When n is even, $y = x^{-n} = \dfrac{1}{x^n} > 0$.

When n is odd, $y = x^{-n} = \dfrac{1}{x^n} > 0$ when $x > 0$ and $y = x^{-n} = \dfrac{1}{x^n} < 0$ when $x < 0$.

Common error: Confusing $y = x^{-n} = \dfrac{1}{x^n}$ with $y = x^{1/n}$ (which we – the authors – did twice while composing this darn question!)

Practice: Graph $y = x^{-3} = \dfrac{1}{x^3}$ and $y = x^{-4} = \dfrac{1}{x^4}$ on the same set of axes.

Answer:

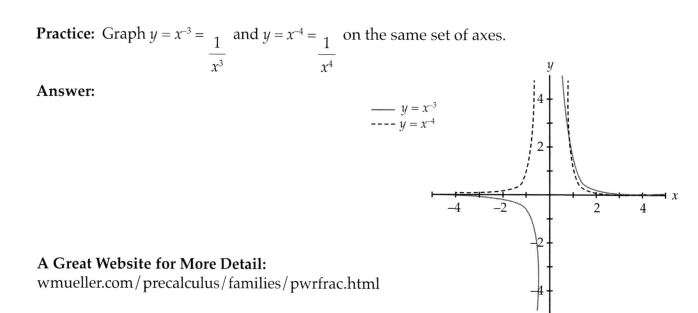

A Great Website for More Detail:
wmueller.com/precalculus/families/pwrfrac.html

5) Graphing $y = x^{-1/n} = \dfrac{1}{x^{1/n}}$, $n \in \mathbf{N}$

Problem: Graph the curves $y = \dfrac{1}{x^{1/2}}$ and $y = \dfrac{1}{x^{1/3}}$

Solution:

$\cdots\cdots\cdots$ $y = \dfrac{1}{x^{1/2}}$

$----$ $y = \dfrac{1}{x^{1/3}}$

Notes: First, we know $x \neq 0$. When n is even, $y = \dfrac{1}{x^{1/n}} > 0$.
When n is odd, $y = \dfrac{1}{x^{1/n}} > 0$ and $y = \dfrac{1}{x^{1/n}}$ is < 0 when $x < 0$.

Common error: Confusing $y = x^{-n} = \dfrac{1}{x^n}$ with $y = \dfrac{1}{x^{1/n}}$

Practice: Graph $y = \dfrac{1}{x^{1/4}}$ and $y = \dfrac{1}{x^{1/5}}$ on the same set of axes.

Answer:

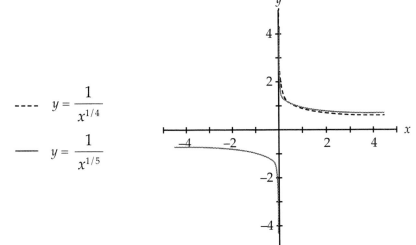

$----$ $y = \dfrac{1}{x^{1/4}}$

$———$ $y = \dfrac{1}{x^{1/5}}$

A Great Website for More Detail: wmueller.com/precalculus/families/pwrfrac.html

6) Transformations (New Graphs from a Given Graph)

Problem: Given $y = f(x) = x^2$, graph and describe each of the following relative to f:
(i) $y = f(x) + 2$ (ii) $y = f(x) - 2$ (iii) $y = f(x + 2)$ (iv) $y = f(x - 2)$
(v) $y = f(2x)$ (vi) $y = 2f(x)$

Solution:
(i) Shifts f up 2 units.   (ii) Shifts f down 2 units.   (iii) Shifts f left 2 units.

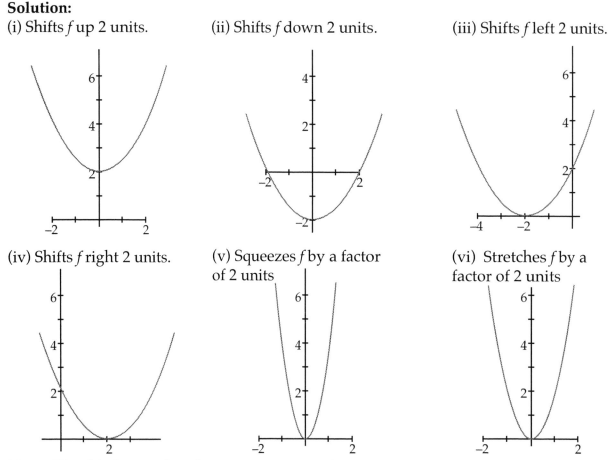

(iv) Shifts f right 2 units.   (v) Squeezes f by a factor of 2 units   (vi) Stretches f by a factor of 2 units

Note: Use this for graphs of every variety–trig, logs, exponents, more!

Common error: Thinking $f(x + 2)$ shifts 2 units to the right because we added 2.

Practice: Given $y = f(x) = \sin(x)$, describe each of the following relative to f.
(i) $y = f(x + \pi)$ (ii) $y = f(\pi x)$ (iii) $y = f(x) + \pi$
(iv) $y = f(x - \pi)$ (v) $y = \pi f(x)$ (vi) $y = f(x) - \pi$

Answer: (i) Shifts f left π units.   (ii) Squeezes f by a factor of π units.
(iii) Shifts f up π units. (iv) Shifts f right π units.   (v) Stretches f by a factor of π units.
(vi) Shifts f down π units.

A Great Website for More Detail:
people.hofstra.edu/Stefan_Waner/calctopic1/scaledgraph.html

ANSWERS
Additional practice
and web references

Part VI: Solving Equations

Part VI: Solving Equations and Inequalities

1) Solving Linear Equations in One Variable

Problem: Solve each of the following equations:

(i) $4x + 20 = 2 - 5x$ (ii) $8(x - 4) + x = 6(x - 5) - (1 - x)$ (iii) $\dfrac{x}{3} - \dfrac{2x}{5} + \dfrac{1}{30} = \dfrac{7}{10}$

Solution: (i) $4x + 20 = 2 - 5x$ $\quad$ | Bring x terms to the left side and constants to the right. | $\quad\Leftrightarrow\quad$ $9x = -18$ $\quad$ | Divide both sides by the coefficient of x. | $\quad\Leftrightarrow\quad$ $x = -2$

(ii) $8(x - 4) + x = 6(x - 5) - (1 - x)$ $\quad$ | Expand and simplify each side. | $\quad\Leftrightarrow\quad$ $9x - 32 = 7x - 31$

| Bring x terms to the left side and constants to the right. | $\quad\Leftrightarrow\quad$ $2x = 1$ $\quad$ | Divide both sides by the coefficient of x. | $\quad\Leftrightarrow\quad$ $x = \dfrac{1}{2}$

(iii) $\dfrac{x}{3} - \dfrac{2x}{5} + \dfrac{1}{30} = \dfrac{7}{10}$ $\quad$ | Multiply the left side and right side by the **least common multiple** of 3, 5 and 10, which is 30. | $\quad\Leftrightarrow\quad$ $10x - 12x + 1 = 21$

| Bring x terms to the left side and constants to the right. | $\quad\Leftrightarrow\quad$ $-2x = 20$ $\quad$ | Divide both sides by the coefficient of x. | $\quad\Leftrightarrow\quad$ $x = -10$

Note: Here is the cardinal rule of high school math:

WYDTOSYDTTO! $\equiv$ WHAT YOU DO TO ONE SIDE YOU DO THE OTHER!

Common error: Messing up one of the cardinal rules of math: "What you do to one side, you do to the other!"

Practice: Solve each of the following equations:

(i) $7(4x - 5) = 8(3x - 5) + 9$ (ii) $\dfrac{x - 1}{2} + \dfrac{2x + 1}{5} = 6$

Answer: (i) $x = 1$ (ii) $x = 7$

A Great Website for More Details:
wtamu.edu/academic/anns/mps/math/mathlab/col_algebra/col_alg_tut14_lineareq.htm

2) Solving Linear Inequalities

Problem: Solve the following inequalities:

(i) $4x - 5 \leq 2x + 9$ (ii) $2x + 7 > 5x - 1$ (iii) $x - 4 < x + 6$

Solution: (i) $4x - 5 \leq 2x + 9$

Bring x terms to the left side and constants to the right.

$\Leftrightarrow \quad 2x \leq 14$

Divide both sides by the coefficient of x.

$\Leftrightarrow \quad x \leq 7 \quad \therefore \ x \in [7, \infty)$

Bring x terms to the left side and constants to the right.

(ii) $2x + 7 > 5x - 1 \quad \Leftrightarrow \quad -3x > -8$

Divide both sides by the coefficient of x.	When you multiply or divide an inequality by a $-$, the direction of the inequality reverses!

$\Leftrightarrow \qquad x \qquad < \qquad \dfrac{3}{8} \quad \therefore \ x \in \left(-\infty, \dfrac{3}{8}\right)$

(iii) $x - 4 < x + 6 \Leftrightarrow 0 < 10$ This is always true! $\therefore x \in \mathbf{R}$

Note: Solve linear inequalities just as you would a linear equation, except when you multiply or divide by a "$-$". Then you must change "$<$" to "$>$" and vice-versa!

Common error: Not paying attention to the note!

Practice: Solve the following inequalities: (i) $5(x - 3) \leq 3(x - 4)$ (ii) $4x - 3 < -7 + 4x$

Answer: (i) $x \in \left(-\infty, \dfrac{3}{2}\right)$ (ii) No solution (ie., ϕ)

A Great Website for More Details:
purplemath.com/modules/ineqlin.htm

3) Solving Two Linear Equations in Two Variables

Problem: Solve for x and y: (E1 and E2 refer to equation 1 and equation 2.)

(i) E1: $x + 2y = -1$ (ii) E1: $x - 2y = 6$ (iii) E1: $x - 2y = 6$

   E2: $5x - 2y = 7$ E2: $3x - 6y = 18$ E2: $3x - 6y = 3$

Solution: (i) Add E1 + E2: $6x = 6 \Leftrightarrow x = 1$ Substitute $x = 1$ into E1:

$1 + 2y = -1 \Leftrightarrow 2y = -2 \Leftrightarrow y = -1$ $\therefore$ The solution is $(x, y) = (1, -1)$

(ii) Multiply $3 \times$ E1 = E3: $3x - 6y = 18$

Subtract E3 – E2: $0 = 0$ This is ALWAYS true! From E1: $x = 6 + 2y$

$\therefore$ Solutions are $\{ (6 + 2y, y) \mid y \in \mathbf{R}\}$

(iii) Multiply $3 \times$ E1 = E3: $3x - 6y = 18$

Subtract E3 – E2 = E4: $0 = 15$ This is NEVER true!

$\therefore$ There are no solutions.

Notes: There are LOTS of ways to solve these problems. Each equation represents a line. In (i), the lines are non-parallel and meet in a single point. In (ii), the lines are "coincident". In (iii), they are parallel and never meet.

Common error: Making an arithmetic mistake when forming a new equation from the given equations.

Practice: Solve $3x - 2y = 7$ and $2x - 5y = 12$

Answer: $(x, y) = (1, -2)$

A Great Website for More Details:
wtamu.edu/academic/anns/mps/math/mathlab/col_algebra/col_alg_tut49_systwo.htm

4) Solving Quadratic Equations

Problem: Solve the following quadratic equations:
(i) $x^2 - 5x + 6 = 0$ (ii) $3x^2 - 7x + 2 = 0$

Solution: (i) $x^2 - 5x + 6 = 0 \iff (x - 2)(x - 3) = 0 \iff x = 2$ or $x = 3$

(ii) $3x^2 - 7x + 2 = 0$

Here, it is easiest to use the quadratic formula: $x = \dfrac{-b \pm \sqrt{b^2 - 4ac}}{2a}$

$$x \underset{\substack{a = 3,\, b = -7,\\ c = 2}}{=} \frac{-(-7) \pm \sqrt{(-7)^2 - 4(3)(2)}}{2(3)} = \frac{7 \pm \sqrt{25}}{6}$$

$$\therefore \; x = \frac{12}{6} = 2 \; \text{ or } \; x = \frac{2}{6} = \frac{1}{3}$$

Note: In (ii), if you factored, you would have found $3x^2 - 7x + 2 = (3x - 1)(x - 2) = 0$, which again (thank the math gods) gives $x = \dfrac{1}{3}$ or $x = 2$.

Common error: Factoring incorrectly.

Practice: Solve the quadratic equations: (i) $x^2 - 3x - 10 = 0$ (ii) $6x^2 - 7x - 3 = 0$

Answer: (i) $x = -2, 5$ (ii) $x = -\dfrac{1}{3}, \dfrac{3}{2}$

A Great Website for More Details: purplemath.com/modules/solvquad.htm

5) Solving Equations Involving Square Roots

Problem: Solve the following equations for x:

(i) $\sqrt{x-2} = 5$ (ii) $\sqrt{4-3x} = x + 12$ (iii) $\sqrt{1+2x} - \sqrt{x} = 1$

Solution: (i) $\sqrt{x-2} = 5 \overset{\boxed{\text{Square both sides.}}}{\Rightarrow} x - 2 = 25 \Rightarrow x = 27$

Check: substitute $x = 27$ in the original equation:
Left Side $= \sqrt{27-2} = \sqrt{25} = 5$ Right side $= 5$ $\therefore x = 5$

(ii) $\sqrt{4-3x} = x + 12 \overset{\boxed{\text{Square both sides.}}}{\Rightarrow} 4 - 3x = x^2 + 24x + 144$

$\boxed{\text{Move "everything" to one side and solve.}}$
$\Rightarrow x^2 + 27x + 140 = 0 \Rightarrow (x+20)(x+7) = 0 \Rightarrow x = -20$ or $x = -7$

Check: substitute $x = -20$ in the original equation:

Left side $= \sqrt{4+60} = 8$ Right side $= -20 + 12 = -8$ $\therefore x = -20$ is NOT a solution.

Check: substitute $x = -7$ in the original equation:

Left side $= \sqrt{4+21} = 5$ Right side $= -7 + 12 = 5$ $\therefore x = -7$ is the only solution.

(iii) $\sqrt{1+2x} - \sqrt{x} = 1$ $\overset{\boxed{\text{Isolate the more complicated square root on one side.}}}{\Rightarrow}$ $\sqrt{1+2x} = 1 + \sqrt{x}$ $\overset{\boxed{\text{Square both sides.}}}{\Rightarrow}$ $1 + 2x = 1 + 2\sqrt{x} + x$

$\boxed{\text{Isolate the remaining square root.}}$ $\boxed{\text{Square both sides.}}$ $\boxed{\text{Move "everything" to one side and solve.}}$
$\Rightarrow 2\sqrt{x} = x \Rightarrow 4x = x^2 \Rightarrow x^2 - 4x = x(x-4) = 0 \therefore x = 0$ or $x = 4$.

Check: Exercise for you but both $x = 0$ and $x = 4$ work!

Note: When we square both sides of an equation, we may introduce solutions to the new equation that are **not** solutions to the original. For example, $x = -3 \Rightarrow x^2 = 9$ The new equation also has $x = 3$ as a solution, but this is not a solution of the original equation $x = -3$. By squaring, we introduced an "extraneous" solution.

Common error: Forgetting to do a left side/right side check for extraneous roots.

Practice: Solve the following equations: (i) $\sqrt{x+4} = 6$ (ii) $\sqrt{x-3} = x - 5$.

Answer: (i) $x = 32$ (ii) $x = 7$

A Great Website for More Detail: purplemath.com/modules/solverad.htm

ANSWERS
Additional practice
and web references

Part VII: Graphing Second Order Relations

Part VII: Graphing Second Order Relations

1) The Parabola

Problem: Graph the following parabolas and identify the vertex and the axis of symmetry: (i) $y = x^2$ (ii) $y = 2(x + 1)^2 - 3$

Solution: (i)

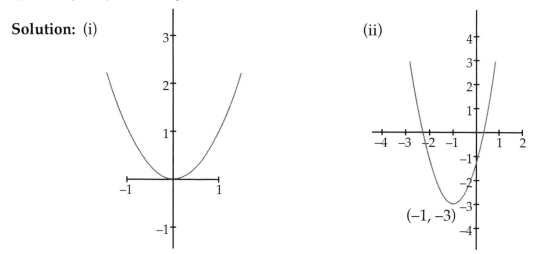

(ii)

Vertex: (0,0); Axis of Symmetry: $x = 0$ Vertex: (−1,−3); Axis of Symmetry: $x = -1$

Note: In $y = a(x - b)^2 + c$, the vertex is (b, c) and the axis of symmetry is $x = b$

When $a > 0$, the parabola opens up; when $a < 0$, the parabola opens down.

Common error: In (ii), identifying the vertex as (1,−3).

Practice: Graph $y = -2(x - 1)^2 + 3$ and identify the vertex and the axis of symmetry.

Answer:

Vertex: (1, 3)
Axis of Symmetry: $x = 1$

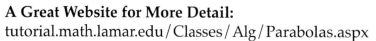

A Great Website for More Detail:
tutorial.math.lamar.edu/Classes/Alg/Parabolas.aspx

2) The Circle

Problem: Graph the following circles and identify the radius and the centre:
(i) $x^2 + y^2 = 4$ (ii) $(x - 1)^2 + (y + 2)^2 = 9$

Solution: (i)   (ii)

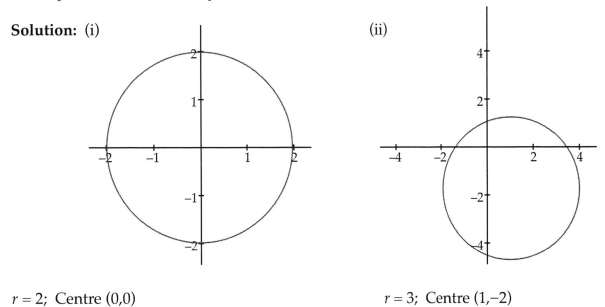

$r = 2$; Centre (0,0) $r = 3$; Centre (1,−2)

Note: In $(x - a)^2 + (y - b)^2 = r^2$, the centre is (a,b) and the radius, of course, is r.

Common error: In (ii), identifying the centre as (−1,2).

Practice: Graph $x^2 + (y + 0.5)^2 = 1$ and identify the radius and the centre.

Answer:

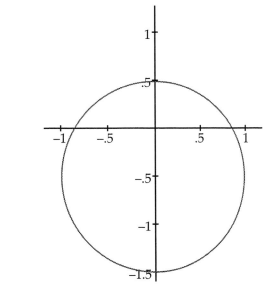

$r = 1$
Centre (0, −0.5)

A Great Website for More Detail:
wtamu.edu/academic/anns/mps/math/mathlab/col_algebra/col_alg_tut29_circles.htm

3) The Ellipse

Problem: Graph the following ellipses and identify the centre and the major and minor axes:

(i) $\dfrac{x^2}{4} + \dfrac{y^2}{9} = 1$ (ii) $\dfrac{(x-1)^2}{4} + \dfrac{(y+2)^2}{9} = 1$

Solution: (i)   (ii)

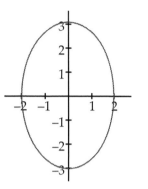
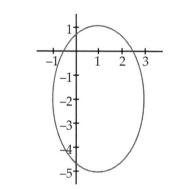

Centre (0,0); Major Axis: y axis (ie., $x = 0$)   Centre (1, –2); Major Axis: $x = 1$
Minor Axis: x axis (ie., $y = 0$)   Minor Axis: $y = -2$

Note: In $\dfrac{x^2}{a^2} + \dfrac{y^2}{b^2} = 1$, the x-intercepts are $\pm a$ and the y-intercepts are $\pm b$

Common error: Confusing the major and minor axes.

Practice: Graph $\dfrac{(x+1)^2}{4} + \dfrac{(y-2)^2}{9} = 1$ and identify the centre and the major and minor axes.

Answer:

Centre: (–1,2)
Major Axis: $x = -1$; Minor Axis: $y = 2$

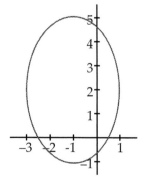

A Great Website for More Detail:
algebralab.org/lessons/lesson.aspx?file=Algebra_conics_ellipse.xml

4) The Hyperbola

Problem: Graph the following hyperbolas and identify the centre and intercepts:

(i) $\dfrac{x^2}{4} - \dfrac{y^2}{9} = 1$ (ii) $\dfrac{y^2}{9} - \dfrac{x^2}{4} = 1$

Solution: (i)

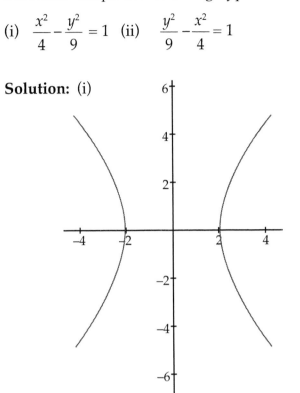

(ii)

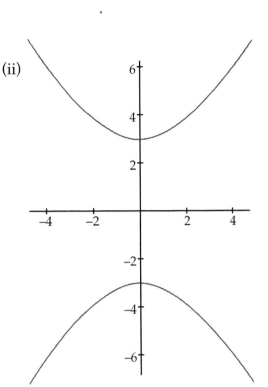

Centre (0,0); x-intercepts $= \pm 2$; no y-intercepts

Centre (0,0); no x-intercepts; y-intercepts $= \pm 3$

Note: $\dfrac{x^2}{a^2} - \dfrac{y^2}{b^2} = 1$ is a hyperbola opening on the x-axis.

$\dfrac{y^2}{b^2} - \dfrac{x^2}{a^2} = 1$ is a hyperbola opening on the y-axis.

Common Error: Putting the intercepts on the wrong axis.

Practice: Graph (i) $x^2 - y^2 = 1$ (ii) $y^2 - x^2 = 1$.

Answer: (i)

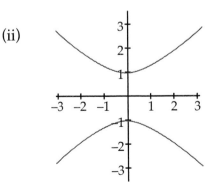

(ii)

A Great Website for More Detail:
tutorial.math.lamar.edu/Classes/Alg/Hyperbolas.aspx

ANSWERS
Additional practice
and web references

Part VIII: Trigonometry

Part VIII: Trigonometry

1) Angles in Standard Position: Degree Measure

Problem: (i) Draw in standard position the following angles:
(a) 30° (b) 225° (c) −80° (d) −190° (e) 390°

(ii) Give all angles, in degrees, which are "co-terminal" with (a) 30° (b) −190°.

Solution:
(i)

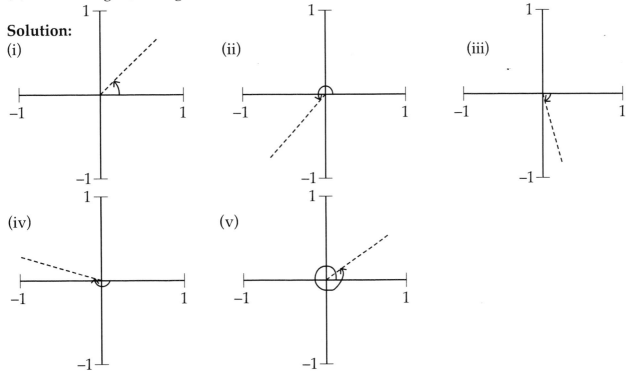

(ii) (a) All angles co-terminal with 30° are $30° + 360°k$, where $k \in \mathbf{Z}$.
(b) All angles co-terminal with −190° are $-190° + 360°k$, where $k \in \mathbf{Z}$.

Note: Co-terminal angles differ by multiples of 360°.

Common error: Regarding co-terminal angles as equal. 30° and 390° are different angles whose terminal arms, in standard position, are the same. Here is an analogy: Consider the function $f(x) = x^2$: $f(-3) = f(3) = 9$ but that doesn't make $-3 = 3$.

Practice: (i) In what quadrant is the terminal arm of −660°?
(ii) Give all the angles in degree measure co-terminal with −660°.

Answer: (i) First quadrant, since −660° is co-terminal with 60°. (ii) $-660° + 360°k$, where $k \in \mathbf{Z}$.

A Great Website for More Details: themathpage.com/aTrig/measure-angles.htm#stand

2) The Meaning of π

Problem: In the circle below with radius *r*, you can "fit" three and a "little portion more" of a radius around half the circle.

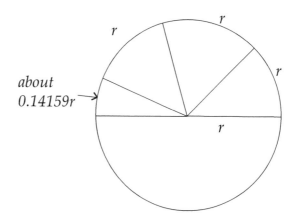

We give a name to this number of radii. (i) The name is _____.

(ii) So the length of a half circle is given by _____.

(iii) This is why the circumference is given by _____.

Solution: (i) π (ii) πr (iii) $2\pi r$

Note: The most commonly used approximations for π are $\dfrac{22}{7}$ or 3.14.

An approximation to π, accurate to 7 decimals, is 3.1415926.

Common problem (as opposed to error): Not having a clue that π is the number of times the radius of a circle wraps around a semi-circle!

Practice: The number π plays a role in the calculation of certain areas. For example, the surface area of a sphere of radius r is $4\pi r^2$ (4 times the area of a circle of radius *r*). Find the surface area of a sphere of radius 3.

Answer: 36π units^2

A Great Website for More Detail: en.wikipedia.org/wiki/Pi

3) Angles in Standard Position: Radian Measure

Problems: 1) Draw in standard position the following angles:

(i) $\dfrac{\pi}{6}$ (ii) $\dfrac{5\pi}{4}$ (iii) $-\dfrac{4\pi}{9}$

2) Give all angles, in radians, which are "co-terminal" with $\pi/6$ radians.

Solutions: 1)(i)   (ii)   (iii)

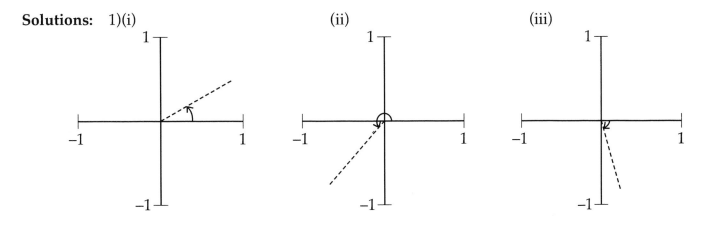

2) All angles co-terminal with $\dfrac{\pi}{6}$ are $\dfrac{\pi}{6} + 2\pi k$, where $k \in \mathbf{Z}$.

Note: 1 radian is about 57°. This is just what you should expect since a little more than 3.14 radians, that is, π radians equals about $3.14 \times 57° \doteq 180°$.

Common error: The formulas for arc length and area of a sector of a circle are

$r\theta$ and $\dfrac{1}{2}r^2\theta$, respectively. Students sometimes substitute θ in degree measure. **It must be in radians!**

Practice: (i) In what quadrant is the terminal arm of $-\dfrac{7\pi}{4}$?

(ii) Give all the angles in degree measure co-terminal with $-\dfrac{7\pi}{4}$.

Answer: (i) First quadrant, since $-\dfrac{7\pi}{4}$ is co-terminal with $\dfrac{\pi}{4}$.

(ii) $-\dfrac{7\pi}{4} + 2\pi k$, where $k \in \mathbf{Z}$.

A Great Website for More Details: themathpage.com/aTrig/radian-measure.htm

4) Degrees to Radians

Problem: Express each of the following in radian measure:

(i) 25° (ii) −150° (iii) 1060° (Remember 180 degrees = π radians.)

Solution: (i) $25° = \dfrac{\pi}{180°} \times 25° = \dfrac{5\pi}{36}$ radians

(ii) $-150° = \dfrac{\pi}{180°} \times (-150°) = -\dfrac{5\pi}{6}$ radians

(iii) $1060° = \dfrac{\pi}{180°} \times (1060°) = \dfrac{53\pi}{9}$ radians

Note: $1° = \dfrac{\pi}{180}$ radians $\therefore x° = x\left(\dfrac{\pi}{180}\right)$ radians

Common misinterpretation: We always use the degree symbol when measuring with degrees but we don't always say radians when using radian measure. So, when measuring angles, "2°", "2 radians", and "2" mean, respectively, "2 degrees", "2 radians", and "2 **radians**", (which is about 115 degrees!).

Practice: Express each of the following in radian measure:

(i) −210° (ii) 300° (iii) 240°

Answer: (i) $\dfrac{-7\pi}{6}$ (ii) $\dfrac{5\pi}{3}$ (iii) $\dfrac{4\pi}{3}$

A Great Website for More Detail:
teacherschoice.com.au/Maths_Library/Angles/Angles.htm

5) Radians to Degrees

Problem: Express each of the following radian measures in degrees:

(i) $\dfrac{4\pi}{9}$ (ii) $\dfrac{2}{5}$ (iii) $-\dfrac{7\pi}{6}$ (iv) $-\dfrac{4\pi}{3}$ (Remember π radians = 180 degrees.)

Solution: (i) $\dfrac{4\pi}{9}$ radians $= \dfrac{4\pi}{9} \times \dfrac{180°}{\pi} = 80°$

(ii) $\dfrac{2}{5}$ radians $= \dfrac{2}{5} \times \dfrac{180°}{\pi} = \dfrac{72}{\pi} = 22.92°$

(iii) $-\dfrac{7\pi}{6}$ radians $= -\dfrac{7\pi}{6} \times \dfrac{180°}{\pi} = -210°$

(iv) $-\dfrac{4\pi}{3}$ radians $= -\dfrac{4\pi}{3} \times \dfrac{180°}{\pi} = -240°$

Note: 1 radian $= \dfrac{180°}{\pi} \doteq 57.3°$ $\therefore x$ radians $= x\left(\dfrac{180°}{\pi}\right)$

Common misinterpretation: We always use the degree symbol when measuring with degrees but we don't always say radians when using radian measure. So, when measuring angles, "2°", "2 radians", and "2" mean, respectively, "2 degrees", "2 radians", and "2 **radians**", (which is about 115 degrees!)

Practice: Express each of the following radian measures in degrees to the nearest degree.

(i) $\dfrac{4}{3}$ (ii) $\dfrac{3\pi}{4}$ (iii) $\dfrac{7\pi}{10}$ (iv) $\dfrac{1}{4}$

Answer: (i) 76°   (ii) 135°   (iii) 126°   (iv) 14°

A Great Website for More Details:
teacherschoice.com.au/Maths_Library/Angles/Angles.htm

6) Relating Angles in Standard Position in Quadrants One and Two

(Related angles: angles whose trig ratios have the same magnitude but differ in sign according to the CAST RULE.)

Problem: (i) Below, the second quadrant angle 170° is drawn in standard position. Find and illustrate the related first quadrant angle, using the interval (0,90°).

(ii) Below, the first quadrant angle $\frac{\pi}{6}$ is drawn in standard position. Find and illustrate the related second quadrant angle, using the interval $\left(\frac{\pi}{2}, \pi\right)$.

Solution: (i) First quad. angle = $180° - 170° = 10°$ (ii) Second quad. angle = $\pi - \frac{\pi}{6} = \frac{5\pi}{6}$

Notes: If θ satisfies $\frac{\pi}{2} < \theta < \pi$, the corresponding first quadrant angle using $\left(0, \frac{\pi}{2}\right)$ is $\pi - \theta$.

If θ satisfies $0 < \theta < \frac{\pi}{2}$, the corresponding second quadrant angle using $\left(\frac{\pi}{2}, \pi\right)$ is still $\pi - \theta$!

Common error: Confusing the second quadrant angle in standard position 170° with the 10° angle that 170° makes with the negative x axis.

Practice: (i) Find the first quadrant angle relatives of (a) 150°   (b) 2π/3
(ii) Find the second quadrant relatives of (a) 75°   (b) π/9.

Answer: (i)(a) 30°   (b) π/3 (ii)(a) 105°   (b) 8π/9

A Great Website for More Detail: oakroadsystems.com/twt/refangle.htm

7) Relating Angles in Standard Position in Quadrants One and Three

(Related angles: angles whose trig ratios have the same magnitude but differ in sign according to the CAST RULE.)

Problem: (i) Below, the third quadrant angle 190° is drawn in standard position. Find and illustrate its related first quadrant angle, using the interval (0,90°).

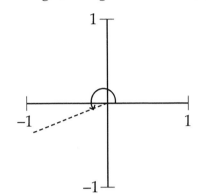

(ii) Below, the first quadrant angle $\frac{\pi}{6}$ is drawn in standard position. Find and illustrate the related second quadrant angle, using the interval $\left(\pi, \frac{3\pi}{2}\right)$.

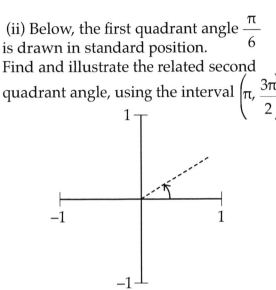

Solution: (i) First quad. angle = 190° − 180° =10°   (ii) Third quad. angle = $\pi + \frac{\pi}{6} = \frac{7\pi}{6}$

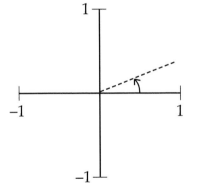

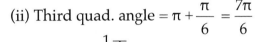

Notes: If θ satisfies $\pi < \theta < \frac{3\pi}{2}$, the corresponding first quadrant angle using $\left(0, \frac{\pi}{2}\right)$ is $\theta - \pi$.

If θ satisfies $0 < \theta < \frac{\pi}{2}$, the corresponding third quadrant angle using $\left(\pi, \frac{3\pi}{2}\right)$ is $\pi + \theta$!

Common error: Confusing the third quadrant angle in standard position 190° with the 10° angle that 190° angle makes with the negative x axis.

Practice: (i) Find the first quadrant angle relatives of (a) 250° (b) $7\pi/6$
(ii) Find the third quadrant relatives of (a) 75° (b) $\pi/9$

Answer: (i)(a) 70° (b) $\pi/6$ (ii)(a) 255° (b) $10\pi/9$

A Great Website for More Detail: oakroadsystems.com/twt/refangle.htm

8) Relating Angles in Standard Position in Quadrants One and Four

(Related angles: angles whose trig ratios have the same magnitude but differ in sign according to the CAST RULE.)

Problem: (i) Below, the fourth quadrant angle −10° is drawn in standard position. Find and illustrate its related first quadrant angle using the interval (0,90°).

(ii) Below, the first quadrant angle $\frac{\pi}{6}$ is drawn in standard position. Find and illustrate its related fourth quadrant angle, using the interval $\left(-\frac{\pi}{2},0\right)$.

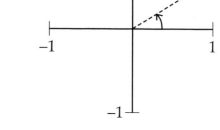

Solution: (i) First quad. angle = −(−10°) = 10°

(ii) Fourth quad. angle = $-\frac{\pi}{6}$

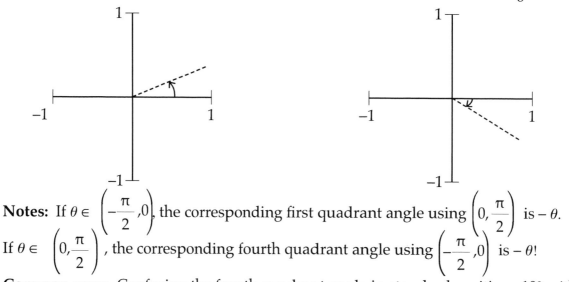

Notes: If $\theta \in \left(-\frac{\pi}{2},0\right)$, the corresponding first quadrant angle using $\left(0,\frac{\pi}{2}\right)$ is − θ.

If $\theta \in \left(0,\frac{\pi}{2}\right)$, the corresponding fourth quadrant angle using $\left(-\frac{\pi}{2},0\right)$ is − θ!

Common error: Confusing the fourth quadrant angle in standard position −10° with the 10° angle that −10° makes with the positive *x* axis.

Practice: (i) Find the **first** quadrant angle relatives of (a) −80° (b) −5π/12.
(ii) Find the **fourth** quadrant relatives of (a) 65° (b) π/3.

Answers: (i)(a) 80° (b) 5π/12 (ii)(a) −65° (b) −π/3
A Great Website for More Detail: oakroadsystems.com/twt/refangle.htm

9) Relating an Angle in Standard Position to its "Relatives" in the other Quadrants

(Related angles: angles whose trig ratios have the same magnitude but differ in sign according to the CAST RULE.)

Quadrant	Angle
1	Degrees: $0° < \theta < 90°$ or Radians: $0 < \theta < \dfrac{\pi}{2}$
2	Degrees: $90° < \theta < 180°$ or Radians: $\dfrac{\pi}{2} < \theta < \pi$
3	Degrees: $180° < \theta < 270°$ or Radians: $\pi < \theta < \dfrac{3\pi}{2}$
4	Degrees: $-90° < \theta < 0°$ or Radians: $-\dfrac{\pi}{2} < \theta < 0$

Problem: State the standard position "relatives" of
(i) 50° (ii) 170° (iii) 250° (iv) − 60° in each of the other quadrants.

Solution: Q ≡ Quadrant
(i) Q2: 130°; Q3: 230°; Q4: − 50° (ii) Q1: 10°; Q3: 190°; Q4: −10°
(iii) Q1: 70°; Q2: 110°; Q4: − 70° (iv) Q1: 60°; Q2: 120°; Q3: 240°

Note: Remember that every angle has **LOTS** of names. Here, we were careful to specify which name we wanted to find.

Common error: You name the angle whose relatives you are trying to find using a range other than specified. For example, you might give 350° when, for the questions on this page, the answer would be –10°.

Practice: State the relatives of (i) $\dfrac{\pi}{4}$ (ii) $\dfrac{9\pi}{7}$ (iii) $\dfrac{13\pi}{10}$ (iv) $-\dfrac{\pi}{5}$

Answer:

(i) Q2: $\dfrac{3\pi}{4}$; Q3: $\dfrac{5\pi}{4}$; Q4: $-\dfrac{\pi}{4}$ (ii) Q1: $\dfrac{2\pi}{7}$; Q3: $\dfrac{5\pi}{7}$; Q4: $-\dfrac{2\pi}{7}$

(iii) Q1: $\dfrac{3\pi}{10}$; Q2: $\dfrac{7\pi}{10}$; Q4: $-\dfrac{3\pi}{10}$ (iv) Q1: $\dfrac{\pi}{5}$; Q2: $\dfrac{4\pi}{5}$; Q3: $\dfrac{6\pi}{5}$

A Great Website for More Detail: oakroadsystems.com/twt/refangle.htm

10) Trigonometric Ratios in Right Triangles: SOHCAHTOA

Problem:

From the triangle, identify all
six trigonometric ratios for θ

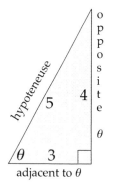

Solution:

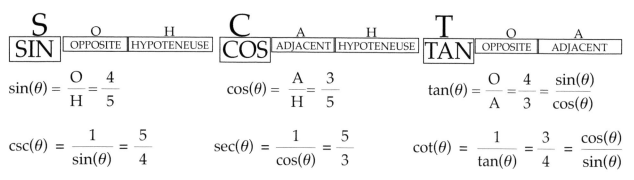

$$\sin(\theta) = \frac{O}{H} = \frac{4}{5}$$

$$\cos(\theta) = \frac{A}{H} = \frac{3}{5}$$

$$\tan(\theta) = \frac{O}{A} = \frac{4}{3} = \frac{\sin(\theta)}{\cos(\theta)}$$

$$\csc(\theta) = \frac{1}{\sin(\theta)} = \frac{5}{4}$$

$$\sec(\theta) = \frac{1}{\cos(\theta)} = \frac{5}{3}$$

$$\cot(\theta) = \frac{1}{\tan(\theta)} = \frac{3}{4} = \frac{\cos(\theta)}{\sin(\theta)}$$

Note: These definitions only work if $0 < \theta < \dfrac{\pi}{2} = 90°$. For other angles, we get the trig values from its "related" first quadrant angle and the CAST RULE.

(Related angles: angles whose trig ratios have the same magnitude but differ in sign according to the CAST RULE.)

Common error: Mixing up the opposite and adjacent sides.

Practice: From the triangle, identify
all six trigonometric ratios for θ.

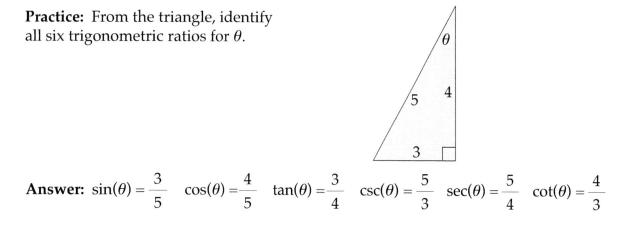

Answer: $\sin(\theta) = \dfrac{3}{5}$ $\quad \cos(\theta) = \dfrac{4}{5}$ $\quad \tan(\theta) = \dfrac{3}{4}$ $\quad \csc(\theta) = \dfrac{5}{3}$ $\quad \sec(\theta) = \dfrac{5}{4}$ $\quad \cot(\theta) = \dfrac{4}{3}$

A Great Website for More Detail: mathwords.com/s/sohcahtoa.htm

11) Trigonometric Ratios Using the Circle: Part I

Problem: Let θ be an angle which is **not** between 0° and 90°.
By drawing the angle in standard position and letting it puncture the unit circle $x^2 + y^2 = 1$ at a point (x, y), find the sin, cos, and tan of θ.

Solution: Draw the angle in standard position. (The illustrated θ satisfies $90° < \theta < 180°$.) It will **puncture** the unit circle, centre the origin, at a point (x, y).

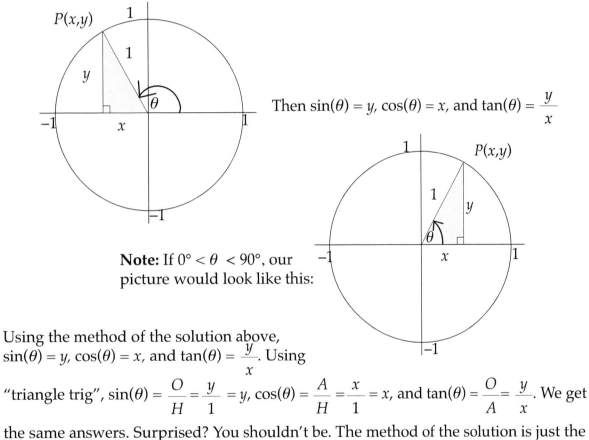

Then $\sin(\theta) = y$, $\cos(\theta) = x$, and $\tan(\theta) = \dfrac{y}{x}$

Note: If $0° < \theta < 90°$, our picture would look like this:

Using the method of the solution above, $\sin(\theta) = y$, $\cos(\theta) = x$, and $\tan(\theta) = \dfrac{y}{x}$. Using "triangle trig", $\sin(\theta) = \dfrac{O}{H} = \dfrac{y}{1} = y$, $\cos(\theta) = \dfrac{A}{H} = \dfrac{x}{1} = x$, and $\tan(\theta) = \dfrac{O}{A} = \dfrac{y}{x}$. We get the same answers. Surprised? You shouldn't be. The method of the solution is just the LOGICAL extension of trig to angles not between 0° and 90°.

Common error: x and y will be either + or − depending on the quadrant where θ punctures the circle. Getting the "signs" wrong (and therefore often the "sin's" wrong!) is common!

Practice: If θ punctures the circle in the fourth quadrant, what are the signs of x and y and how does this relate to the CAST RULE?

Answer: x is + and y is −. The sin and tan will be − while cos will be +. This accounts for the **C** in the CAST RULE.

A Great Website for more Detail: themathpage.com/aTrig/unit-circle.htm#quad

12) Trigonometric Ratios Using the Circle: Part II

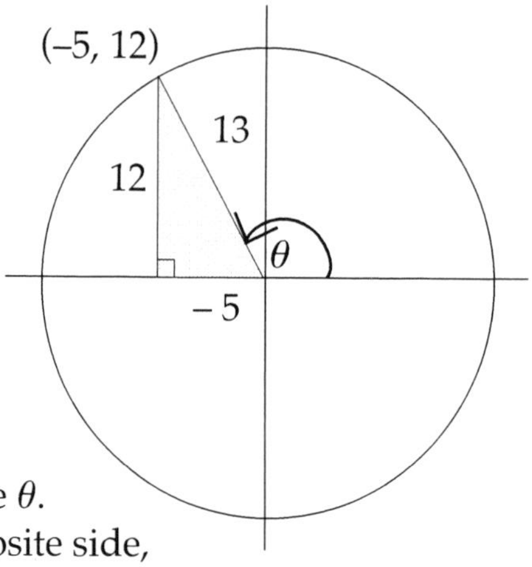

Problem: In the diagram, θ, where $90° < \theta < 180°$, is a second quadrant angle in standard position whose terminal side punctures the circle (centered at the origin, with radius 13) at the point P(-5,12). State the six trigonometric ratios of θ.

Solution:
The second quadrant triangle with sides -5, 12, and 13 is the right triangle associated with the second quadrant angle θ. Using this triangle with -5 as the adjacent side, 12 as the opposite side, and 13 as the hypotenuse, we have

$$\sin(\theta) = \frac{O}{H} = \frac{12}{13} \qquad \cos(\theta) = \frac{A}{H} = \frac{-5}{13} = -\frac{5}{13} \qquad \tan(\theta) = \frac{O}{A} = \frac{12}{-5} = -\frac{12}{5}$$

$$\csc(\theta) = \frac{13}{12} \qquad \sec(\theta) = -\frac{13}{5} \qquad \cot(\theta) = -\frac{5}{12}$$

Notes: If $\alpha = \theta + 360°$, then α will have the same second quadrant associated triangle as θ and so will have the same trig ratios. This is true for **all** angles of the form $\theta + 360k°$, where $k \in \mathbf{Z}$. Note that of the three ratios sin, cos, and tan, only sin is positive. This is where the S in the CAST RULE (for the Sin in the second quadrant) comes from. **Another Note:** We could have used the circle with radius 1 and the associated triangle

with sides $\frac{12}{13}$ and $-\frac{5}{13}$. The resulting triangle is **similar** to the one we used above and

so the trig ratios would be the same.
And One More Note: Every angle in standard position will have an "associated" right triangle. Take the end point (a,b) on the terminal arm (the point where θ, in standard position, punctures the circle) and draw a perpendicular to the x axis. There is your triangle. The adjacent side is a, the opposite side is b, and the hypotenuse is $\sqrt{a^2 + b^2}$.

Common error: Mislabeling the signs on the co-ordinates of the point where θ punctures the circle: for example, mislabeling P as (5,12) instead of (-5,12).

Practice: Suppose θ, where $\pi < \theta < \frac{3\pi}{2}$ punctures the circle, centred at the origin and with radius 5, at the point $(-4,-3)$. State the sin, cos, and tan of θ. (Hint: hypotenuse = 5)

Answer: $\sin(\theta) = -\frac{3}{5} \qquad \cos(\theta) = -\frac{4}{5} \qquad \tan(\theta) = \frac{3}{4}$

A Great Website for more Detail: themathpage.com/aTrig/unit-circle.htm#quad

13) Trigonometric Ratios for the 45°, 45°, 90° Triangle

Problem: Find the values of all the six trigonometric ratios of $45° = \dfrac{\pi}{4}$.

Solution: Let $\triangle ABC$ be an isosceles right triangle with $\angle B = 90°$ and $AB = CB = 1$. Then $\angle A = \angle C = 45°$ and $AC = \sqrt{2}$.

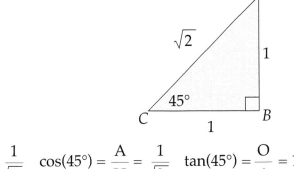

$$\sin(45°) = \frac{O}{H} = \frac{1}{\sqrt{2}} \quad \cos(45°) = \frac{A}{H} = \frac{1}{\sqrt{2}} \quad \tan(45°) = \frac{O}{A} = 1$$

$$\csc(45°) = \sqrt{2} \qquad \sec(45°) = \sqrt{2} \qquad \cot(45°) = 1$$

Note: We set $AB = CB = 1$. If we had set $AB = CB = 3$, then we would have $AC = 3\sqrt{2}$

Then, $\sin(45°) = \dfrac{3}{3\sqrt{2}} = \dfrac{1}{\sqrt{2}}$, which is the same as before. In other words, the sides

stay in proportion to one another (similar triangles!) and the ratios are unchanged.

Common error: Determining the hypotenuse $AC = 2$.

Practice: Find the values of the six trigonometric ratios of the second quadrant angle 135°. (Hint: 135° has related first quadrant angle 180° − 135° = 45°. Use the ratios for 45° and the CAST RULE.)

Answer: $\sin(135°) = \dfrac{1}{\sqrt{2}} \qquad\qquad \cos(135°) = -\dfrac{1}{\sqrt{2}} \qquad\qquad \tan(135°) = -1$

$$\csc(135°) = \sqrt{2} \qquad\qquad \sec(135°) = -\sqrt{2} \qquad\qquad \cot(135°) = -1$$

A Great Website for more Detail:

hyperad.com/tutoring/math/trig/Trigonometric%20Functions%20of%20Common%20Angles.html

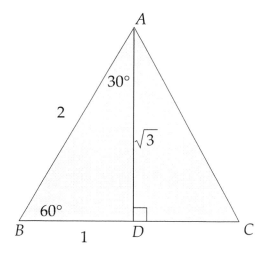

14) Trigonometric Ratios for the 30°, 60°, 90° Triangle

Problem: Find the values of all the six

trigonometric ratios of $60° = \dfrac{\pi}{3}$ and $30° = \dfrac{\pi}{6}$.

Solution: Let $\triangle\, ABC$ be an equilateral
triangle with $AB = CB = AC = 2$. Then
$\angle A = \angle B = \angle C = 60°$. Drop a perpendicular
from A to meet BC at D. $\therefore \triangle\, ADB \cong \triangle\, ADC$ and so
$\angle BAD = \angle CAD = 30°$ and $BD = CD = 1$.

Finally, using Pythagoras, $AD = \sqrt{2^2 - 1^2} = \sqrt{3}$

$$\sin(60°) = \frac{O}{H} = \frac{\sqrt{3}}{2} \qquad\qquad \cos(60°) = \frac{A}{H} = \frac{1}{2} \qquad\qquad \tan(60°) = \frac{O}{A} = \sqrt{3}$$

$$\csc(60°) = \frac{2}{\sqrt{3}} \qquad\qquad \sec(60°) = 2 \qquad\qquad \cot(60°) = \frac{1}{\sqrt{3}}$$

$$\sin(30°) = \frac{O}{H} = \frac{1}{2} \qquad\qquad \cos(30°) = \frac{A}{H} = \frac{\sqrt{3}}{2} \qquad\qquad \tan(30°) = \frac{O}{A} = \frac{1}{\sqrt{3}}$$

$$\csc(30°) = 2 \qquad\qquad \sec(30°) = \frac{2}{\sqrt{3}} \qquad\qquad \cot(30°) = \sqrt{3}$$

Note: The adjacent side for 60°, 1, is the opposite side for 30°.
The opposite side for 60°, $\sqrt{3}$, is the adjacent side for 30°. This **always**
happens with "complementary" angles. The adjacent for θ is the opposite for $90° - \theta$.

Common error: Forgetting where to put the 1 and the $\sqrt{3}$. Everybody remembers that 2
is the hypoteneuse.

Practice: Find the sin, cos and tan of the third quadrant angle 210°. (Hint: 210° has related first quadrant angle $210° - 180° = 30°$. Use the ratios for 30° and the CAST RULE.)

Answer: $\sin(210°) = -\dfrac{1}{2} \qquad \cos(210°) = -\dfrac{\sqrt{3}}{2} \qquad \tan(210°) = \dfrac{1}{\sqrt{3}}$

A Great Website for more Detail:
hyperad.com/tutoring/math/trig/Trigonometric%20Functions%20of%20Common
%20Angles.html

15) Trigonometric Ratios for the 0°, 90°, 180°, 270°; ie., Trig Ratios For Angles ON the Axes

First please re-read "11) Trigonometric Ratios Using the Circle: Part I".

Problem: Find the six trig ratios for $0° = 0$ radians and $90° = \dfrac{\pi}{2}$ radians.

Solution: $\theta = 0°$ punctures the unit circle at $(1,0)$ and $90°$ punctures this circle at $(0,1)$.

$0°$

$\sin(0°) = 0$	$\cos(0°) = 1$	$\tan(0°) = \dfrac{0}{1} = 0$
$\csc(0°)$ is undefined	$\sec(0°) = 1$	$\cot(0°)$ is undefined

$90°$

$\sin(90°) = 1$	$\cos(90°) = 0$	$\tan(90°)$ is undefined
$\csc(90°) = 1$	$\sec(90°)$ is undefined	$\cot(90°) = 0$

Notes: Division by zero is undefined and that is why there are undefined trig ratios for angles that puncture the axes, that is, where one of the co-ordinates is 0. Also, the ratios for $360k°$, for $k \in \mathbf{Z}$, are the same as for $0°$. The ratios for $90° + 360k°$, for $k \in \mathbf{Z}$, are the same as for $90°$.

Common error: $\sin(0°)=1$, $\cos(0°)=0$, etc.

Practice: Find the sin, cos and tan of (i) $\dfrac{7\pi}{2}$ (ii) -3π

Hint: $\dfrac{7\pi}{2}$ is co-terminal with $-\dfrac{\pi}{2}$ and punctures the circle at $(0, -1)$;

-3π is co-terminal with π and punctures the circle at $(-1,0)$.

Answer: (i) $\sin \dfrac{7\pi}{2} = -1$ $\cos \dfrac{7\pi}{2} = 0$ $\tan \dfrac{7\pi}{2}$ is undefined

(ii) $\sin(-3\pi) = 0$ $\cos(-3\pi) = -1$ $\tan(-3\pi) = 0$

A Great Website for More Detail:
jwbales.home.mindspring.com/precal/part4/part4.2.html

16) CAST RULE

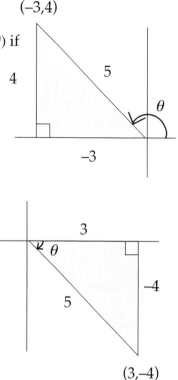

Problems: 1) Given $\tan(\theta) = -4/3$, find the values of $\sin(\theta)$ and $\cos(\theta)$ if
(i) θ is a second quadrant angle.
(ii) θ is a fourth quadrant angle.
2) Why can't θ be a first or third quadrant angle?

Solutions: (i) If $\tan(\theta) = -\dfrac{4}{3}$, and θ is a second quadrant

angle, then $x < 0$ and $y > 0$. Draw θ with a terminal point
$(-3,4)$ and drop a perpendicular to the x axis to get the
"associated" right triangle. (In the picture, we assumed

$\dfrac{\pi}{2} < \theta < \pi$ but θ could have satisfied, for example,

$-\dfrac{3\pi}{2} < \theta < -\pi$.) Then $\sin(\theta) = \dfrac{4}{5}$ and $\cos(\theta) = -\dfrac{3}{5}$.

(ii) If $\tan(\theta) = -\dfrac{4}{3}$, and θ is a fourth quadrant angle,

then $x > 0$ and $y < 0$. Draw θ with terminal point $(3,-4)$
and drop a perpendicular to the x axis to get the "associated"

right triangle. Then $\sin(\theta) = -\dfrac{4}{5}$ and $\cos(\theta) = \dfrac{3}{5}$.

2) tan is **positive** in quadrants 1 and 3 so we can't have $\tan(\theta) = -\dfrac{4}{3}$!

Note:

QUADRANT 4	QUADRANT 1	QUADRANT 2	QUADRANT 3
C	A	S	T
COS	ALL	SIN	TAN

Common error: Given $\tan(\theta) = -\dfrac{4}{3}$ in the second quadrant, setting $x = 3$ and $y = -4$.

Practice: Given $\cos(\theta) = -\dfrac{2}{3}$ and θ is a third quadrant angle, find $\sin(\theta)$ and $\tan(\theta)$.

Answer: $\sin(\theta) = -\dfrac{\sqrt{5}}{3}$ $\tan(\theta) = \dfrac{\sqrt{5}}{2}$

A Great Website for More Detail:
worsleyschool.net / science / files / cast / castdiagram.html

 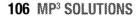

17) Sine Law: Find an Angle

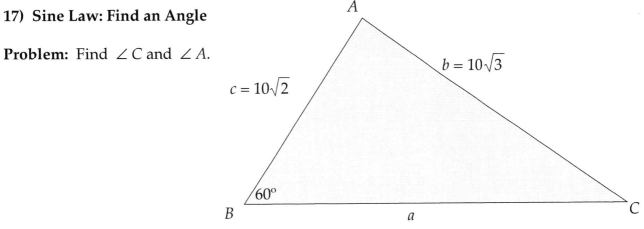

Problem: Find $\angle C$ and $\angle A$.

$c = 10\sqrt{2}$

$b = 10\sqrt{3}$

$60°$

Solution: By the Sine Law,

$$\frac{\sin(C)}{c} = \frac{\sin(B)}{b}$$

$b = 10\sqrt{3}$
$c = 10\sqrt{2}$

$\sin(60°) = \dfrac{\sqrt{3}}{2}$

$$\therefore \sin(C) = \frac{c}{b}\sin(B) = \frac{10\sqrt{2}}{10\sqrt{3}} \times \frac{\sqrt{3}}{2} = \frac{1}{\sqrt{2}}$$

We know the trig ratios for 45°.

$$\therefore \angle C = 45° \text{ and } \angle A = 180° - (60° + 45°) = 75°$$

Notes: $\sin(C) = \dfrac{1}{\sqrt{2}} \Rightarrow C = 45°$ or $135°$. (Sin is + in both the first and second quadrants!)

The angles in $\triangle ABC$ must sum to $180°$ and so $C = 135°$ is too big. However, sometimes, there are two solutions!

Another Note: This works when you have two sides and an angle which is **not the contained angle**. If you have the contained angle, you need the Cosine Law.

Common error: $a\sin(A) = b\sin(B)$

Practice: In $\triangle ABC$, $c = AB = 10$, $b = AC = 12$ and $\angle B = 60°$. Find $\angle C$ and $\angle A$ accurate to one decimal place. (Hint: to solve $\sin(C) = x$ on most calculators, use " 2nd Function sin ".)

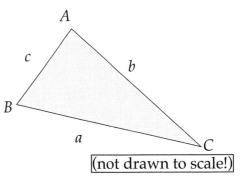

(not drawn to scale!)

Answers: $\angle C \doteq 46.2°$ $\angle A \doteq 73.8°$

A Great Website for More Detail:
ilearn.senecac.on.ca / learningobjects / MathConcepts / SineLaw / main.htm

18) Sine Law: Find a Side

Problem: Find (i) the exact value of a and (ii) c accurate to two decimals.

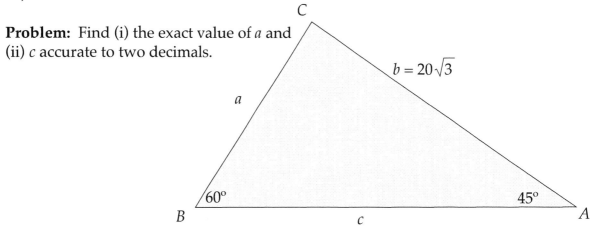

Solution: $\angle C = 180° - (60° + 45°) = 75°$

By the Sine Law,

$$a = \sin(A) \times \frac{b}{\sin(B)} = \sin(45°) \times \frac{20\sqrt{3} \times 2}{\sqrt{3}} = \frac{1}{\sqrt{2}} \times 40 = 20\sqrt{2}$$

$$c = \sin(C) \times \frac{b}{\sin(B)} = \sin(75°) \times \frac{20\sqrt{3} \times 2}{\sqrt{3}} \doteq 0.9659 \times 40 \doteq 38.64$$

Note: The Sine Law contains four quantities: two angles in a triangle and the two sides opposite these angles. The Sine Law is useful when you have three of these four quantities.

Common error: $a\sin(A) = b\sin(B)$

Practice: In $\triangle ABC$, $\angle A = 60°$, $\angle B = 45°$ and side $b = AC = 36$. Find the lengths of sides $a = BC$ and $c = AB$ accurate to two decimal places.

Answer: $a \doteq 44.10$ $c \doteq 49.18$

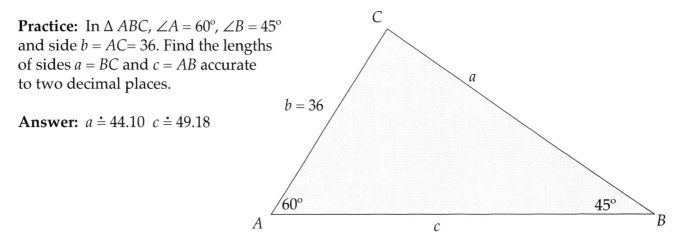

A Great Website for More Detail:
ilearn.senecac.on.ca/learningobjects/MathConcepts/SineLaw/main.htm

19) Cosine Law: Find an Angle

Problem: In $\triangle ABC$, use the Cosine Law to find $\angle B$ to the nearest degree.

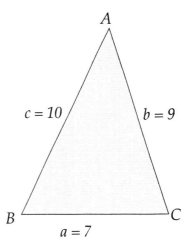

Solution: Using the Cosine Law:

$$\cos(B) = \frac{a^2 + c^2 - b^2}{2ac} = \frac{49 + 100 - 81}{2(7)(10)} = \frac{68}{140} = \frac{17}{35}$$

Make sure your calculator is in "DEGREE" mode!

On most calculators:
17/35 = 2ND FUNCTION cos =

$\therefore \; B \doteq 60.94°$

Note: Make sure your calculator is in degree mode. To find an angle using the Cosine Law you need to know the lengths of all the three sides of the triangle.

Common error: Using "RADIAN" mode when the answer is required in degrees or vice-versa. The answer here in radians is 1.06. Remember that 1 radian is about 57°. So, 1.06 radians is about 61°.

Practice: In $\triangle ABC$, we have $a = BC = 2$, $b = AC = 3$ and $c = AB = 4$. Find $\angle C$ to the nearest degree.

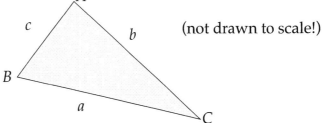

(not drawn to scale!)

Answer: $\angle C = 105°$.

A Great Website for More Detail:
ilearn.senecac.on.ca/learningobjects/MathConcepts/CosineLaw/main.htm

20) Cosine Law: Find a Side

Problem: In $\triangle ABC$, use the Cosine Law to find $c = AB$ correct to two decimal places.

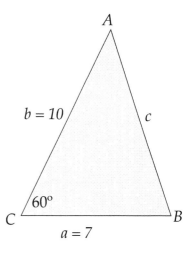

Solution: Using the Cosine Law:

$c^2 = a^2 + b^2 - 2 \cdot a \cdot b \cdot \cos(C)$ $\boxed{a = 7, b = 10, \angle C = 60°}$ $= 49 + 100 - 140(0.5) = 79$

$\therefore c \doteq 8.89$

Note: To find a side using the Cosine Law, you need two sides and the contained angle. If you have a "non-contained" angle, use the Sine Law to find the contained angle and then use the Cosine Law or the Sine Law to find the required side.

Common error: $c^2 = a^2 + b^2 + 2ab\cos(C)$

Practice: In $\triangle ABC$, we have $b = AC = 6$, $c = AB = 4$ and $\angle A = 35°$. Find $a = BC$, correct to two decimals.

Answer: $a \doteq 3.56$

A Great Website for More Detail:
ilearn.senecac.on.ca/learningobjects/MathConcepts/CosineLaw/main.htm

21) The Graphs of the Sin, Cos, and Tan Functions

Problem: Graph for $-2\pi \leq \theta \leq 2\pi$: (i) $y = \sin(\theta)$ (ii) $y = \cos(\theta)$ (iii) $y = \tan(\theta)$

Solution: (i) $y = \sin(\theta)$ (ii) $y = \cos(\theta)$ (iii) $y = \tan(\theta)$

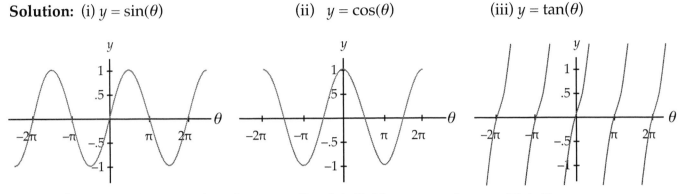

Note: The range of $y = \sin(\theta)$ and $y = \cos(\theta)$ is [-1,1]. The range of $y = \tan(\theta)$ is **R**.

Common error: Confusing, for example, the point $\left(\dfrac{\pi}{4}, \dfrac{1}{\sqrt{2}}\right)$, which is on the graph of both $y = \sin(\theta)$ and $y = \cos(\theta)$, with the terminal point $\left(\dfrac{1}{\sqrt{2}}, \dfrac{1}{\sqrt{2}}\right)$ of $\theta = \dfrac{\pi}{4}$, on the circle $x^2 + y^2 = 1$. On this circle,

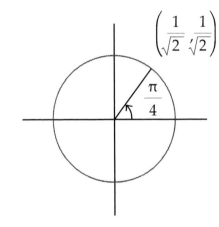

for each angle θ, we puncture the circle in a point (x, y) and we define $x = \cos(\theta)$ and $y = \sin(\theta)$. The values of cos and sin lead to the graphs above.

Practice: Name all values of θ where (i) $\sin(\theta) = 0$ (ii) $\cos(\theta) = 0$ (iii) $\tan(\theta) = 0$.

Answer: (i) $k\pi, k \in \mathbf{Z}$ (ii) $\dfrac{\pi}{2} + k\pi, k \in \mathbf{Z}$ (iii) $k\pi, k \in \mathbf{Z}$

A Great Website for More Detail:
library.thinkquest.org/20991/alg2/trig.html#graph

22) Period of the Sin, Cos, and Tan Functions

Problems: State the period of each of the following:

1)(i) $y = \sin(x)$ (ii) $y = \sin(2x)$ (iii) $y = \sin\left(\dfrac{x}{2}\right)$

2)(i) $y = \cos(x)$ (ii) $y = \cos(3x)$ (iii) $y = \cos\left(\dfrac{x}{3}\right)$

3)(i) $y = \tan(x)$ (ii) $y = \tan(4x)$ (iii) $y = \tan\left(\dfrac{\pi}{2}x\right)$

Solutions: 1)(i) 2π (ii) $\dfrac{2\pi}{2} = \pi$ (iii) $\dfrac{2\pi}{\left(\dfrac{1}{2}\right)} = 4\pi$

2)(i) 2π (ii) $\dfrac{2\pi}{3}$ (iii) $\dfrac{2\pi}{\left(\dfrac{1}{3}\right)} = 6\pi$

3)(i) π (ii) $\dfrac{\pi}{4}$ (iii) $\dfrac{\pi}{\left(\dfrac{\pi}{2}\right)} = 2$

Note: If the function $y = f(x)$ has period P, then $y = f(kx)$ has period $\dfrac{P}{k}$.

Common error: The period of $y = \tan(2x)$ is $\dfrac{2\pi}{2} = \pi$.

Practice: Because the period of $y = \sin(x)$ is 2π, we know that for any integer k, $\sin(x + 2k\pi) =$ _____.

Answer: $\sin(x)$

A Great Website for More Detail: mathsrevision.net/alevel/pages.php?page=36

23) The Graphs of the Csc, Sec, and Cot Functions

Problem: Graph for $-2\pi \leq \theta \leq 2\pi$:   (i) $y = \csc(\theta)$ (ii) $y = \sec(\theta)$ (iii) $y = \cot(\theta)$

Solution: (i) $y = \csc(\theta)$ (ii) $y = \sec(\theta)$ ·   (iii) $y = \cot(\theta)$

Note: The range of $y = \csc(\theta)$ and $y = \sec(\theta)$ is $(-\infty, -1] \cup [1, \infty)$. The range of $y = \cot(\theta)$ is $\mathbf{R}$.

Common error: Completely misunderstanding why the range of csc is from +1 up and from −1 down: when $\sin(\theta)$ is inside $(-1,1)$, then $\dfrac{1}{\sin(\theta)} = \csc(\theta)$ is outside $(-1,1)$. When $\sin(\theta) = \pm 1$, so does $\csc(\theta)$.

Practice: Name all values of θ where (i) $\csc(\theta)$ (ii) $\sec(\theta)$ (iii) $\cot(\theta)$ is undefined.

Answer: (i) $k\pi, k \in \mathbf{Z}$ (ii) $\dfrac{\pi}{2} + k\pi, k \in \mathbf{Z}$ (iii) $k\pi, k \in \mathbf{Z}$

A Great Website for More Detail:
regentsprep.org/Regents/math/algtrig/ATT7/othergraphs.htm

24) Trig Formulas That You Should Know

Problem: Complete the following formulas:

(i) $\sin^2(\theta) + \cos^2(\theta) = $ _____ ; $1 + \tan^2(\theta) = $ _____

(ii) $\sin(-\theta) = $ _____ ; $\cos(-\theta) = $ _____

(iii) $\sin\left(\dfrac{\pi}{2} - \theta\right) = $ _____ ; $\cos\left(\dfrac{\pi}{2} - \theta\right) = $ _____

(iv) $\sin(A \pm B) = $ _____ ; $\cos(A \pm B) = $ _____

(v) $\sin(2A) = $ _____ ; $\cos(2A) = $ _____

Solution: (i) $\sin^2(\theta) + \cos^2(\theta) = \boxed{1}$; $1 + \tan^2(\theta) = \boxed{\sec^2(\theta)}$

(ii) $\sin(-\theta) = \boxed{-\sin(\theta)}$; $\cos(-\theta) = \boxed{\cos(\theta)}$

(iii) $\sin\left(\dfrac{\pi}{2} - \theta\right) = \boxed{\cos(\theta)}$; $\cos\left(\dfrac{\pi}{2} - \theta\right) = \boxed{\sin(\theta)}$

(iv) $\sin(A \pm B) = \boxed{\sin(A)\cos(B) \pm \cos(A)\sin(B)}$;

$\cos(A \pm B) = \boxed{\cos(A)\cos(B) \mp \sin(A)\sin(B)}$

(v) $\sin(2A) = \boxed{2\sin(A)\cos(A)}$; $\cos(2A) = \boxed{\cos^2(A) - \sin^2(A)}$

Note: These formulas are used over and over and over again. Learn them.
Common error: $\cos(A \pm B) = \boxed{\cos(A)\cos(B) \pm \sin(A)\sin(B)}$

Practice: Complete the formulas:

(i) $\cot^2(\theta) + 1 = $ _____ (ii) $\tan(-\theta) = $ _____ (iii) $\tan\left(\dfrac{\pi}{2} - \theta\right) = $ _____

(iv) $\tan(A + B) = $ _____ (v) $\tan(2A) = $ _____

Answer: (i) $\boxed{\csc^2(\theta)}$ (ii) $\boxed{-\tan(\theta)}$ (iii) $\boxed{\cot(\theta)}$

(iv) $\tan(A + B) = \boxed{\dfrac{\tan(A) + \tan(B)}{1 - \tan(A)\tan(B)}}$ (v) $\tan(2A) = \boxed{\dfrac{2\tan(A)}{1 - \tan^2(A)}}$

A Great Website for More Detail: analyzemath.com / trigonometry / trigonometric_formulas.html

ANSWERS

Additional practice and web references

Part IX: Logs and Exponents

Part IX: Logs and Exponents

1) Exponents

Problems:

1) Evaluate: (i) 2^3 (ii) $\left(\dfrac{3}{5}\right)^3$ (iii) 4^{-2} (iv) 10^0 (v) $\left(\dfrac{1}{0.01}\right)^{-3}$

2) Simplify: (i) $\dfrac{x^5 x^4}{x^7}$ (ii) w^{-1} (iii) $\dfrac{1}{w^{-3}}$ (iv) $\left(z^{2/3}\right)^{10}$ (v) $\left(\dfrac{a^7 b^3}{c^2}\right)^5$

Solution: 1) (i) $2^3 = 2 \times 2 \times 2 = 8$ (ii) $\left(\dfrac{3}{5}\right)^3 = \dfrac{3^3}{5^3} = \dfrac{27}{125}$ (iii) $4^{-2} = \dfrac{1}{4^2} = \dfrac{1}{16}$

(iv) $10^0 = 1$ (v) $\left(\dfrac{1}{0.01}\right)^{-3} = (0.01)^3 = (10^{-2})^3 = 10^{-6} = \dfrac{1}{1000000}$

2) $\dfrac{x^5 x^4}{x^7} = x^{5+4-7} = x^2$ (ii) $w^{-1} = \dfrac{1}{w}$ (iii) $\dfrac{1}{w^{-3}} = w^3$ (iv) $\left(z^{2/3}\right)^{10} = z^{20/3}$ (v) $\left(\dfrac{a^7 b^3}{c^2}\right)^5 = \dfrac{a^{35} b^{15}}{c^{10}}$

Notes: Multiplication is a short form for repeated addition: $7 + 7 + 7 + 7 = 7 \times 4$
Exponentiation is a short form for repeated multiplication: $7 \times 7 \times 7 \times 7 = 7^4$

Common error: $(3^5)^4 = 3^9$, ie., adding exponents when you should be multiplying.

Practice: 1) Evaluate (i) 2^5 (ii) $\left(\dfrac{2}{3}\right)^{-2}$ (iii) 10^0 (iv) 0^0

2) Simplify: (i) $\dfrac{x^5 y^2}{x^{11} y^{-5}}$ (ii) h^{-1} (iii) $\dfrac{1}{h^{-1}}$ (iv) $\left(\dfrac{ab^{-2}}{c^{-3}}\right)^5$

Answers: 1)(i) 32 (ii) $\dfrac{9}{4}$ (iii) 1 (iv) Does not exist. "0^0" is undefined.

2)(i) $\dfrac{y^7}{x^6}$ (ii) $\dfrac{1}{h}$ (iii) h (iv) $\dfrac{a^5 c^{15}}{b^{10}}$

A Great Website for More Detail: purplemath.com/modules/exponent.htm

2) Logarithms (Log means FIND THE EXPONENT!)

Problems: 1) Evaluate: (i) $\log_2 8$ (ii) $\log_2\left(\dfrac{1}{8}\right)$ (iii) $\log_3 1$ (iv) $\log_5 5$

(v) $\log 10000$ (vi) $\ln(e^7)$*

2) Expand using log properties: $\ln\left(\dfrac{x^3 y^{1/2}}{z^4}\right)$

3) Change $\log_5 7$ to log with base 3, then with base 10, and finally with base e.

> * "e" and "ln" refer to the "natural logarithm". If you have not taken calculus, you may be totally unfamiliar with e. If so, treat it as a constant just as you would, for example, a.

Solutions: 1) (i) $\log_2 8 = 3$ (ii) $\log_2\left(\dfrac{1}{8}\right) = -3$ (iii) $\log_3 1 = 0$ (iv) $\log_5 5 = 1$

(v) $\log 10000 = 4$ (vi) $\ln(e^7) = 7$

2) $\ln\left(\dfrac{x^3 y^{1/2}}{z^4}\right) = \ln(x^3) + \ln(y^{1/2}) - \ln(z^4) = 3\ln x + \dfrac{1}{2}\ln y - 4\ln z$

3) $\log_5 7 = \dfrac{\log_3 7}{\log_3 5} = \dfrac{\log 7}{\log 5} = \dfrac{\ln 7}{\ln 5}$

Notes: "Log" means "Find the exponent!"
"log" with no base is short for $\log_{10}$ and ln is short for $\log_e$.

Common error: Many students confuse $(\log_2 8)^3 = \log_2 8 \times \log_2 8 \times \log_2 8 = 3^3 = 27$
with $\log_2(8^3) = \log_2(8 \times 8 \times 8) = \log_2(2^9) = 9$.

Practice: 1) Evaluate: (i) $\log_3 81$ (ii) $\log_5\left(\dfrac{1}{125}\right)$ (iii) $\log_{0.1} 1$ (iv) $\log_2 0$

(v) $\log\dfrac{1}{10000}$ (vi) $\ln\left(\dfrac{1}{e^7}\right)$

2) Expand using log properties: $\ln(x^3 + y^{1/2} - z^4)$

3) Change $\log_5 7$ to log with base 7.

Answers: 1) (i) 4 (ii) 3 (iii) 0 (iv) does not exist (v) –4 (vi) –7

2) You can't expand this at all. 3) $\dfrac{1}{\log_7 5}$

A Great Website for More Detail: purplemath.com/modules/logrules.htm

3) Exponential Graphs

Problem: (i) Graph the exponential functions $y = 2^x$ and $y = 3^x$ on the same set of axes.

(ii) Graph the exponential functions $y = 2^{-x} = \dfrac{1}{2^x}$ and $y = 3^{-x} = \dfrac{1}{3^x}$ on the same set of axes.

Solution: (i)

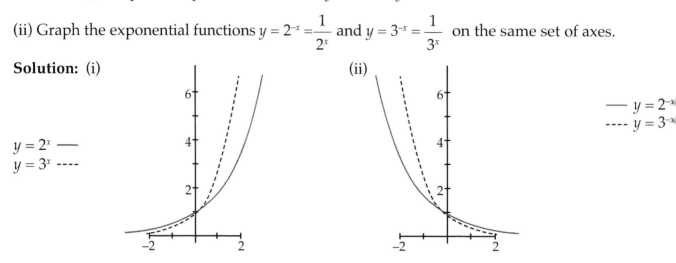

$y = 2^x$ ——
$y = 3^x$ ----

Notes: Because the base is bigger (and greater than 1), $y = 3^x$ goes up faster than $y = 2^x$

when $x > 0$ and y approaches 0 faster when $x \to -\infty$. Also, $y = a^x$ is **always** +!

$y = a^{-x}$ and $y = a^x$ are mirror images in the y axis.

Common error: Drawing the graph of $y = 2^x$ so that it appears to cross the x axis when $x \to -\infty$.

Practice: Graph $y = 10^x$ and $y = 10^{-x}$ on the same set of axes.

Answer:

$y = 10^x$ ——
$y = 10^{-x}$ ----

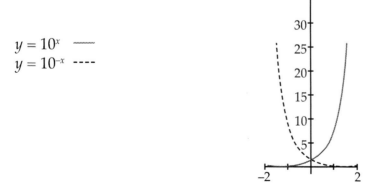

A Great Website for More Detail: purplemath.com/modules/graphexp.htm

4) Logarithmic Graphs

Problem: Graph the functions $y = \log_2(x)$ and $y = \log_3(x)$ on the same set of axes.

Solution:

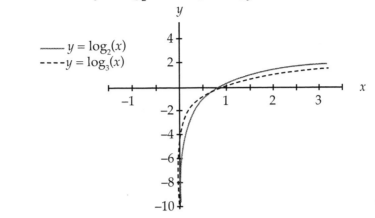

Notes: Because the base is bigger (and greater than 1), $y = \log_3(x)$ goes up more slowly than $y = \log_2(x)$ when $x > 1$ and y approaches $-\infty$ faster when $x \to 0^+$. The domain of $y = \log_a(x)$ is $(0,\infty)$!

Common error: Drawing the graph of $y = \log_2(x)$ so that it appears to cross the y axis when $x \to 0^+$. Even worse, this means you are allowing $x < 0$ into $y = \log_2(x)$. NO!

Practice: Graph $y = \log_2(x)$ and $y = \log_2\left(\dfrac{1}{x}\right)$ on the same set of axes.

Hint: $\log_2\left(\dfrac{1}{x}\right) = \log_2(x^{-1}) = -\log_2(x)$

Answer:

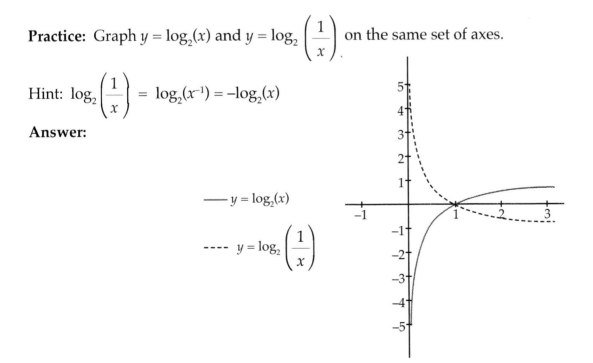

A Great Website for More Detail: purplemath.com/modules/graphlog.htm

5) Exponents to Logarithms and Vice-Versa

Problems: 1) Change to a log equation: (i) $32 = 2^5$ (ii) $y = 10^x$ (iii) $y = 4x^k$ (Use base 10.)

2) Change to an exponential equation: (i) $\log_3 81 = 4$ (ii) $y = \log_5 x$

Solutions: 1)(i) $32 = 2^5 \Leftrightarrow 5 = \log_2 32$ (ii) $y = 10^x \Leftrightarrow x = \log_{10}(y) = \log(y)$

(iii) $y = 4x^k$ $\boxed{\text{Take log of each side.}}$ $\Leftrightarrow$ $\log(y) = \log(4x^k)$ $\boxed{\text{log properties}}$ $= \log(4) + k\log(x)$

2) (i) $\log_3 81 = 4 \Leftrightarrow 81 = 3^4$ (ii) $y = \log_5 x \Leftrightarrow x = 5^y$

Note: $y = a^x \Leftrightarrow x = \log_a(y)$

Common error: $y = 4x^k \Leftrightarrow k = \log_{4x}(y)$

Practice: 1) Change to a log equation: (i) $\dfrac{8}{27} = \left(\dfrac{2}{3}\right)^3$ (ii) $y = 0.5x^{k-1}$ (Use base 10.)

2) Change to an exponential equation: (i) $\log(0.001) = -3$ (ii) $y + 2 = \log_5(3x - 1)$

Answers: 1) (i) $\log_{2/3}\dfrac{8}{27} = 3$ (ii) $\log(y) = \log(.5) + (k-1)\log(x)$

2)(i) $10^{-3} = .001$ (ii) $5^{y+2} = 3x - 1$

A Great Website for More Detail: purplemath.com/modules/logs.htm

6) Using a Calculator to Evaluate Exponents and Logs

Problems: Use a calculator to give answers rounded to two decimal places.

1) (i) $2^{3.3}$ (ii) $5^{1/7}$ (iii) $(-10)^{1/3}$

2) (i) $\log(25)$ (ii) $\ln(25)$*   (iii) $\log_2(25)$

---
*"e" and "ln" refer to the "natural logarithm". If you have not taken calculus, you may be totally unfamiliar with e. If so, treat it as a constant just as you would, for example, a.

Solutions: 1) (i) $2^{3.3}$ Most Calculators: 2.2 y^x 3.3 = $=$ 9.85   (ii) $5 = 5^{1/7}$ Most Calculators: 7 $\sqrt[x]{y}$ 5 = $\doteq$ 1.26

(iii) $(-10)^{1/3}$. Most calculators won't accept a base < 0.

We know the answer should be negative. So

$(-10)^{1/3} = -(10)^{1/3}$ Most Calculators: 3 $\sqrt[x]{x}$ 10 = $\doteq$ -2.15

2)(i) $\log(25)$ Most Calculators: 25 log = $\doteq$ 1.40   (ii) $\ln(25)$ Most Calculators: 25 ln = $\doteq$ 3.22

(iii) $\log_2(25)$ Most Calculators: 25 log / 2 log = $\doteq$ 4.64   OR   Most Calculators: 25 ln / 2 ln = $\doteq$ 4.64

Note: Use the change of base formula when given a base other than 10 or e:

$$\log_a(b) = \frac{\log(b)}{\log(a)} = \frac{\ln(b)}{\ln(a)}$$

Common error: Believing your calculator when it answers INCORRECTLY when you enter -2 y^x 3. Most calculators will report, "error" because they are not programmed to handle a negative base. Of course, $(-2)^3 = -8$.

Practice: Use a calculator to give answers rounded to two decimal places.
1(i) $5^{3.1}$ (ii) $2^{1/5}$ (iii) $(-1001)^{1/3}$
2(i) $\log(6)$ (ii) $\ln(6)$ (iii) $\log_7(6)$

Answers: 1)(i) 146.83   (ii) 1.15   (iii) -10.00
2(i) .78   (ii) 1.79   (iii) .92

A Great Website for More Detail:
geocities.com/Athens/Thebes/5118/scicalc/scicalc.htm

MP³ Notes

MP³ Notes

MP³ Notes